'Eve Croeser presents a compelling case for an emancipatory project for climate justice and ecosocialism. Mobilising Marxist political economy and an ecological neo-Gramscianism, her rigorous, accessible and elegantly written book should help inspire the activism needed to save the biosphere and for ecosocialism to triumph over capitalist barbarism.'

Adrian Budd, Associate Professor, London South Bank University, UK

'Empirically rich and analytically deft in its account of the rise of both the Anthropocene and global capitalism, *Ecosocialism and Climate Justice* probes the pitfalls of mainstream institutional responses to the climate crisis, and points the way towards the socially just alternative that is so urgently needed.'

William K. Carroll, Professor of Sociology and Co-director, the Corporate Mapping Project, University of Victoria, Canada

Ecosocialism and Climate Justice

This book investigates the broader climate movement to contextualise the role played by its climate justice wing, focusing specifically on the theoretical and practical contributions of ecosocialists.

Ecosocialism and Climate Justice provides an account of the shift from the Holocene to the Anthropocene in the context of the global spread of capitalist relations of production. Croeser begins by critically analysing the root causes of anthropogenic climate change and identifies the origins and development of the current climate movement within civil society. She then focuses on the climate justice movement, analysing the ways in which anthropogenic global warming may be challenged in a way that is socially just. Overall, this book provides further insight into the effectiveness of ecosocialist theory and activism in the context of existing global, national and local power relationships.

This book will be of great interest to students and scholars of climate justice, climate politics, critical global political economy studies and environmental activism.

Eve Croeser is a University Associate within the School of Social Sciences at the University of Tasmania, Australia.

Routledge Studies in Environmental Justice

This series is theoretically and geographically broad in scope, seeking to explore the emerging debates, controversies and practical solutions within Environmental Justice, from around the globe. It offers cutting-edge perspectives at both a local and global scale, engaging with topics such as climate justice, water governance, air pollution, waste management, environmental crime, and the various intersections of the field with related disciplines.

The *Routledge Studies in Environmental Justice* series welcomes submissions that combine strong academic theory with practical applications, and as such is relevant to a global readership of students, researchers, policymakers, practitioners and activists.

Ecosocialism and Climate Justice
An Ecological Neo-Gramscian Analysis
Eve Croeser

For more information about this series, please visit: www.routledge.com/Routledge-Studies-in-Environmental-Justice/book-series/EJS

Ecosocialism and Climate Justice

An Ecological Neo-Gramscian Analysis

Eve Croeser

LONDON AND NEW YORK

First published 2021
by Routledge
2 Park Square, Milton Park, Abingdon, Oxon OX14 4RN

and by Routledge
605 Third Avenue, New York, NY 10017

First issued in paperback 2022

Routledge is an imprint of the Taylor & Francis Group, an informa business

Publisher's Note
The publisher has gone to great lengths to ensure the quality of this reprint but points out that some imperfections in the original copies may be apparent.

British Library Cataloguing-in-Publication Data
A catalogue record for this book is available from the British Library

Library of Congress Cataloging-in-Publication Data
Names: Croeser, Eve, author.
Title: Ecosocialism and climate justice : an ecological neo–Gramscian analysis / Eve Croeser.
Description: Abingdon, Oxon ; New York, NY : Routledge, 2021. | Includes bibliographical references and index.
Identifiers: LCCN 2020019130 (print) | LCCN 2020019131 (ebook) | ISBN 9780367894535 (hardback) | ISBN 9781003019268 (ebook)
Subjects: LCSH: Gramsci, Antonio, 1891–1937. | Cox, Robert W., 1926–2018. | Environmental justice. | Ecosocialism. | Climatic changes–Philosophy.
Classification: LCC GE220 .C75 2021 (print) | LCC GE220 (ebook) | DDC 304.2–dc23
LC record available at https://lccn.loc.gov/2020019130
LC ebook record available at https://lccn.loc.gov/2020019131

ISBN 13: 978-0-367-55944-1 (pbk)
ISBN 13: 978-0-367-89453-5 (hbk)
ISBN 13: 978-1-003-01926-8 (ebk)

DOI: 10.4324/9781003019268

Typeset in Bembo
by Wearset Ltd, Boldon, Tyne and Wear

This book is dedicated to 'the 99%', and to all the children and future generations who will suffer the most in the dystopia that threatens us if we do not immediately take effective action on mitigating further anthropogenic climate change.

Contents

List of illustrations   x
Preface   xi
Acknowledgements   xiv
List of abbreviations   xvi

1   The organic crisis of global capitalism in the
    Anthropocene   1

2   Theoretical perspectives for the Anthropocene:
    an ecological neo-Gramscian Method of Historical
    Structures   24

3   Competing ideas: ecosocialist theory   53

4   Institutional responses to a changing biosphere:
    the Intergovernmental Panel on Climate Change and
    the United Nations Framework Convention on
    Climate Change   81

5   The social dynamics of 'climate justice' versus
    'climate action' in the climate movement   110

6   Competing ideas: ecosocialist strategy and tactics
    in the struggle for climate justice   130

7   The biosphere and social forces in a geopolitically
    unstable world beset by organic crisis   163

8   Prospects for climate justice: a research agenda   192

Index   197

Illustrations

Figures

1.1	Socio-economic trends	5
1.2	Earth System trends	6
2.1	Cox's MHS Forces	35
2.2	Cox's MHS Spheres	36
2.3	Sinclair's MHS Forces Redux	37
2.4	Sinclair's MHS Spheres Redux	38
2.5	MHS Forces Redux II	42
4.1	Millennial Northern Hemisphere (NH) temperature reconstruction and instrumental data from AD 1000 to 1999	94
4.2	Socio-economic trends (OECD, BRICS, others)	100
6.1	Ecosocialist strategy and tactics	152
7.1	Earth System tipping elements at risk	169
8.1	Overview of extended case study: an ecological neo-Gramscian concept map of the global political economy of climate change	193

Tables

6.1	Summary of ecosocialist questions regarding strategy and tactics	137
6.2	Reformist versus revolutionary tactics	142

Preface

I wrote this book because we live in exceptional times and I felt compelled to try to contribute to finding solutions to the existential crisis humanity faces as the life support systems we depend on are degraded and destroyed by powerful economic interests focused on short-term material gains. As I complete the final draft of this book, I have been distracted by the avalanche of news articles with titles such as 'Antarctic region Sees Record High Temperature of 18.3°C amid Rising Concern over Melting Ice Sheets' (Australian Broadcasting Corporation, 8 February 2020). In this news article, World Meteorological Organization (WMO) spokesperson Clare Nullis states that 'The amount of ice lost annually from the Antarctic ice sheet increased at least six-fold between 1979 and 2017' and warns that 'Scientists believe global warming has caused so much melting at the south pole that the giant ice sheet is now on course to disintegrate'. Such articles compound the plethora of distressing daily news reports on the unprecedented ferocity of the November 2019–January 2020 Australian bushfires. Despite the death of 33 people, the destruction of what is estimated to be more than 20% of Australia's forests, the deaths of over a billion animals and the appalling scenes of residents and holidaymakers huddled on beaches as their escape routes were cut off by the uncontrollable flames, the Australian government refuses to consider increasing its Paris Agreement greenhouse gas (GHG) emission reduction targets while acknowledging that it understands that these emissions are contributing to anthropogenic global warming (AGW) and climate change.

In addition to regular news updates on the destruction of the Great Barrier Reef, other articles attempt to alert readers to equally important, albeit less visible, environmental changes – such as the death of the giant kelp forests in the Tasman Peninsula (Australian Broadcasting Corporation, 8 February 2020), with reef ecologist Scott Ling's conclusion that 'What we've learned here in Tasmania and the world over is we actually need to throw out the textbooks from back in the day and relearn these ecosystems.' More disturbingly, the slow violence that daily affects millions of poor and disadvantaged people in less economically 'developed' parts of the world, such as the locust infestation that has been decimating many African and Middle Eastern crops since January 2020, and other 'natural disasters' (such as 'extreme' droughts and floods) that are exacerbated by the effects of AGW, is barely mentioned by most news media outlets.

The distress caused by reading such media reports is greatly compounded if one consults the major scientific reports summarising the findings of the most recent research on the damaging effects of AGW, including climate change and biodiversity loss. For more than three decades, the official reports published by scientific bodies such as the WMO and the Intergovernmental Panel on Climate Change have been warning the world's policymakers of the need to mitigate additional AGW in order to try to limit the climate change, ocean acidification, biodiversity loss and general environmental degradation and destruction that is already locked in and must be adapted to. Even publications produced by groups representing extremely wealthy elites, such as the World Economic Forum's *Global Risks Report* (2020), emphasise the urgency of addressing these issues. Despite all these warnings, if one considers the outcomes of what were touted as the 2019 Australian and British 'climate elections', most of the voting public seems to prioritise a variety of issues other than the existential emergency we currently face as a result of AGW.

There are many complex reasons for the evident public complacency in the so-called 'advanced economies' of the world regarding the climate crisis and the health of the environment, including the fact that many people are struggling just to get by and worry about what implications environmental protection and a shift to renewable energy sources will have on their precarious jobs. Such concerns are not only promoted by many politicians and policymakers but are also exacerbated by the rise of the political Right and the attendant politicisation of environmental issues such as the climate change crisis. AGW denialism and/or trivialisation is, moreover, currently popularised through the dissemination of 'fake news' – both online and within some sectors of the official media. In other cases, official media outlets either wittingly or inadvertently fail to generate public concern about these crises. Instead of matching the urgency of the climate crisis with daily headlines focusing on AGW, climate change, environmental destruction and biodiversity loss, they prioritise their reporting on a range of other issues and problems, such as the US–China 'trade war' and various other conflicts that a succession of US administrations have been involved in over the years – the 2020 impeachment proceedings against US President Donald Trump, 'Brexit' and so on.

As I complete the final revisions for this book, however, a more immediate global crisis has been rapidly unfolding: the coronavirus disease 2019 (COVID-19) pandemic. Governments have legislated many measures to try to contain the pandemic, including closing their national borders to foreign citizens, ordering most businesses to shut and restricting their own citizens' movements and activities. The long-term effects of both the pandemic and the measures being taken to try to minimise the number of people who die (either directly because of complications arising from contracting COVID-19 or indirectly because of the widespread and large-scale disruption it is causing to ill-prepared national health services) are currently unknown. It is, however, almost certain that the repercussions of the 2019–2020 pandemic will be profound. It would not be surprising if the current world order changes dramatically after the pandemic; for example, global supply chains have been severely disrupted and it is highly likely that their pre-pandemic configuration will change dramatically once the worst of the crisis is over. It is even conceivable that the COVID-19 pandemic heralds the end of the neoliberal globalisation project discussed in this book, although it is far too early to even begin looking for signals of its demise.

If one considers the pressing issue of climate change, which scientists tell us we must address in the next decade if we are to keep the rise in average global temperatures to 1.5 °C, the COVID-19 pandemic could not have come at a worse time. While mandatory workplace shut-downs and restrictions on population movements have dramatically reduced GHG emissions in the short term, dealing with the pandemic has sidelined all other issues, including the climate crisis, at all levels – globally, nationally, locally and even at the individual level. Another possible repercussion of the pandemic is its probable adverse effect on the production of, and investments in, renewable technologies. Echoing Naomi Klein's observation that it is our great misfortune that the neoliberal globalisation project was gathering speed at precisely the time when climate scientists discovered the severity of AGW, it now seems to be our great misfortune that the pandemic has stalled what little action was happening to address the climate crisis. The situation is, however, fluid, and the only thing we can be certain of right now is that there will be surprising outcomes. Depending on people's reactions, we may emerge from this unfolding catastrophe both stronger and more determined to look after our vital life-support systems and one another. This is where the hope now lies; another world is possible if enough people experience these awful times as a 'wake-up call'.

Acknowledgements

First and foremost, I would like to express my deepest thanks and appreciation to my partner and best friend, John Studholme, who supported me in every way possible throughout the doctoral studies preceding the writing of this book as well as throughout the long process of transforming the thesis into the book. Without John's encouragement, patient care and reassurances, it is unlikely that I would have either embarked on or completed this journey. John's unwavering support took many forms, including participating in many discussions with me that challenged my assumptions and identifying arguments that did not make sense. His valuable feedback helped me present my ideas more clearly and logically, and I thank him profusely for all his support and assistance.

I would also like to express my deep appreciation to Professor Fred Gale, who was the primary supervisor throughout my doctoral studies at the University of Tasmania. Fred very competently, consistently and selflessly guided me through research that eventually led to the writing of this book. Like John, Fred consistently challenged me to question my assumptions and justify my arguments. In doing so, he helped me develop a more sophisticated analysis, and for this I am very grateful. Fred admirably achieved all this while simultaneously ensuring that I had the intellectual space to reach my own conclusions, and all arguments and views presented in this book are entirely my own.

In addition, I would like to thank my daughter, Dr Sky Croeser, who critically reviewed early drafts of the manuscript and, like the rest of my family, motivated me with her unconditional support for this project. As always, my family provided the encouragement and stability I need to succeed in all my endeavours. I am particularly grateful to my parents, who instilled in me the important values that inform this book: a dedication to truth and knowledge, and a deep caring for other people. One of the most important habits my father helped me develop as he listened to radio news broadcasts, every hour on the hour, was that of paying attention to my world. I am very grateful to him for thereby provoking a lifelong interest in current affairs that led, indirectly, to my awareness of contemporary global challenges and hence to the research project culminating in this book.

I would also like the thank my interviewees, who gave so generously of their valuable time to respond to my questions; in particular, I am grateful to Ian Angus for setting aside time for two interviews as well as less formal discussions in between his Australian conference presentations in 2016. I am also very grateful to the two anonymous reviewers for the time they devoted to reviewing my original manuscript, and for all their valuable and insightful feedback, which I used as a guide when working on the revisions. Other individuals I am indebted to and would like to thank for their attention and encouragement throughout the challenges of completing this book include Associate Professor Timothy Sinclair and editorial assistant Matt Shobbrook. Tim kindly reviewed several drafts of an academic paper that informed some of the ideas discussed in this book, providing valuable feedback as well as much-appreciated encouragement. Matt has been very patient with me, and I thank him for all the time he devoted to answering my questions and for his advice and encouragement, which are very much appreciated.

Abbreviations

AGGG	Advisory Group on Greenhouse Gases
APEC	Asia-Pacific Economic Co-operation
AR4	Fourth Assessment Report of the Intergovernmental Panel on Climate Change
AR5	Fifth Assessment Report of the Intergovernmental Panel on Climate Change
AR6	Sixth Assessment Report of the Intergovernmental Panel on Climate Change
ASEAN	Association of Southeast Asian Nations
BLF	Builders Labourers Federation
CAN	Climate Action Network
CAN-I	Climate Action Network International
CBDR	common but differentiated responsibilities
CDM	Clean Development Mechanism
CJN!	Climate Justice Now!
COP	Conference of the Parties
CSO	civil society organisation
EDF	Environmental Defense Fund
EIN	Ecosocialist International Network
ENGO	environmental non-governmental organisation
EPA	Environmental Protection Authority (US)
ETS	emission trading scheme
EU	European Union
FAO	Food and Agricultural Organisation
FAR	First Assessment Report of the Intergovernmental Panel on Climate Change
FDI	foreign direct investment
FoE	Friends of the Earth
FoE-I	Friends of the Earth International
GATT	General Agreement on Tariffs and Trade
GBR	Great Barrier Reef
GCC	Global Climate Coalition
GFC	Global Financial Crisis of 2007/2008

GHG	greenhouse gas
GJM	Global Justice Movement
GPE	global political economy
IAC	InterAgency Council
IBRD	International Bank for Reconstruction and Development
ICSU	International Council for Science
ICT	information and communication technology
ILO	International Labor Organization
IMF	International Monetary Fund
IMS	International Monetary System
INDCs	Intended Nationally Determined Contributions
IO	international organisation
IPCC	Intergovernmental Panel on Climate Change
IR	international relations
IS	Islamic State
ITWF	International Transport Workers' Federation
JI	joint implementation
LULUCF	land use, land use change and forestry
MENA	Middle East and North Africa
MERCOSUR	Common Market of the South
MHS	Method of Historical Structures
MIT	Massachusetts Institute of Technology
MRV	monitoring, reporting and verification
NAFTA	North American Free Trade Agreement
NASA	National Aeronautics and Space Administration
NATO	North Atlantic Treaty Organisation
NGO	non-governmental organisation
OECD	Organisation for Economic Co-operation and Development
OPEC	Organisation of Petroleum Exporting Countries
REDD+	reducing emissions from deforestation and forest degradation
SAR	Second Assessment Report of the Intergovernmental Panel on Climate Change
SCNCC	System Change Not Climate Change
SEI	Stockholm Environment Institute
SMO	social movement organisation
SPM	Summary for Policymakers
SR15	An IPCC Special Report on the impacts of global warming of 1.5°C above pre-industrial levels and related global greenhouse gas emission pathways, in the context of strengthening the global response to the threat of climate change, sustainable development, and efforts to eradicate poverty
TAR	Third Assessment Report of the Intergovernmental Panel on Climate Change
TCC	transnational capitalist class

TNC	transnational corporations
TUED	Trade Unions for Energy Democracy
UN	United Nations
UNCHE	United Nations Conference on the Human Environment
UNCSD	United Nations Conference on Sustainable Development
UNCTAD	United Nations Conference on Trade and Development
UNEP	United Nations Environment Program
UNFCCC	United Nations Framework Convention on Climate Change
UNGA	United Nations General Assembly
UNHCR	United Nations High Commissioner for Refugees
WCC	World Climate Conference
WCED	World Commission on Environment and Development
WEF	World Economic Forum
WG	working group
WHO	World Health Organisation
WMO	World Meteorological Organization
WRI	World Resources Institute
WSF	World Social Forum
WTO	World Trade Organization
WWF	World Wildlife Fund

1 The organic crisis of global capitalism in the Anthropocene

Introduction

The world currently confronts an interlinked ecological, economic, social and political crisis crystallised in the issue of climate change. Despite increasingly urgent warnings from the scientific community that we need to drastically reduce anthropogenic greenhouse gas (GHG) emissions within the next decade if we aim to limit the rise in average global temperatures to 1.5 °C above pre-industrial levels (IPCC, 2018a), emissions continue to rise and there is no sign that they will be 'peaking in the next few years' (UNEP, 2019). This is despite the publication of the Intergovernmental Panel on Climate Change's (IPCC) 2018 *Special Report on Global Warming of 1.5 °C* (SR15), which warns that the damage to ecosystems and socio-economic systems of allowing global temperatures to rise to 2 °C above pre-industrial levels is likely to be much more severe than if the temperature rise is limited to 1.5 °C (IPCC, 2018a).[1] The scenario we face is even worse than this, however; even if governments meet their formal commitments to reducing GHG emissions by the amount they have nominated in the current international climate agreement (the Paris Agreement), SR15 warns that this is likely to 'result in a global warming of about 3 °C by 2100, with warming continuing afterwards' (IPCC, 2018a, p. 4).

What life will be like on a planet that is 3 °C warmer is difficult to imagine; even at the current level of warming, which has raised the global temperature by approximately 1 °C ('with a likely range of between 0.8 °C and 1.2 °C') above the pre-industrial average (IPCC, 2018a, p. 4), the damaging effects of 'extreme weather' events such as the Australian 2019/2020 'megafires' are unprecedented.[2] The ecological and socio-economic crises that such events precipitate demonstrate that, as SR15 explains, while a 1.5 °C temperature rise will be more manageable than a 2 °C increase, it will still pose great dangers to ecosystems and the people who live in them (IPCC, 2018a). From a climate justice perspective, the risks that climate change poses to the least advantaged people compound the challenges they already face as they contend with insecure livelihoods in an increasingly conflict-ridden, unstable and unequal world.

While disappointing, the lack of urgency in policymakers' responses to the IPCC SR15 recommendations is not surprising given that more than two decades of international climate change negotiations have yielded very little in terms of real action. As discussed in Chapter 5, the continuing failure of governments acting alone and through international institutions to effectively address the climate crisis has led to the growth of a distinct climate movement within civil society whose broad aim is to bring about the changes required to mitigate anthropogenic global warming (AGW). Like many other social movements, the climate movement is heterogeneous, and its members have different ideas about how best to address the climate crisis. While some activists call for 'climate action' (which involves doing the technical things necessary to mitigate anthropogenic climate change, such as 'leaving fossil fuels in the ground'), others insist on 'climate justice' and call for 'system change'. The ecosocialists who are the main subjects of the study in this book, and whose theoretical and practical contributions to the climate movement are discussed in Chapters 3 and 6, fall into the latter category. As well as analysing the role of ecosocialists within the climate movement, several other interlinked aims informed the writing of this book.

Main themes and aims

The primary aim of this book is to contribute to efforts to find socially just solutions to the climate crisis that is currently unfolding. The ecosocialists who are the focus of this book have themselves published many excellent books and papers analysing the causes of the anthropogenic climate crisis and proposing socially just solutions to it. This book is not intended to compete with the extensive body of ecosocialist writings; it is, rather, an overview and meta-analysis of some key aspects of ecosocialist thought and practice with the aim of introducing these ideas to readers who may not be aware of them. Another aim of this book is to identify ecosocialist strengths as well as the challenges they face in the hope that this will provoke further thought and debate about these issues both amongst ecosocialists and amongst other readers who are concerned about the social justice issues arising from the interlinked crises we currently face. Ecosocialists contribute many valuable ideas on the issue of climate change that merit a wide audience; for example, their definition of the concept of 'climate justice' is much broader and deeper than liberal definitions of this concept and can fruitfully contribute to the urgent debate we need to engage in regarding how to respond to the human suffering that results from the interlinked ecological, economic, political and social crises we currently face.

While ecosocialist thought and practice is one of the book's main themes, the theoretical lens through which this theme is explored was developed in response to Clive Hamilton's (2017) argument that the advent of the Anthropocene requires new approaches to the way we do research, as well as to what and how we teach. Hamilton's views of the implications of the Anthropocene for the social sciences are discussed in more detail later in this chapter; it suffices to say at this point that his observations motivated the incorporation of a new category, the 'Biosphere', into the standard neo-Gramscian theoretical perspective first developed by Robert Cox (1981, 1983) and later modified by his colleague Timothy Sinclair (2016), so that it is more explicitly ecological. The incorporation of the global political economy as a category *within* the Earth's biosphere is the modest novel theoretical contribution that this book makes to the existing academic literature on the issues of climate change and climate justice.

The original neo-Gramscian perspective and its modifications are described in Chapter 2. The ecological neo-Gramscian perspective is then used as an heuristic device throughout the rest of the book to achieve the book's final goal: to conduct an extensive case study that demonstrates how this theoretical lens can facilitate a critical ecological political economy account of the interconnected crises we face. It is hoped that other analysts will find the ecological neo-Gramscian perspective interesting and, perhaps, even useful in their own work. More generally, the material discussed in this book is intended to stimulate further reflection on, and exchanges of ideas regarding, the pressing issue of how to shift from 'Holocene' to 'Anthropocene' modes of thinking, analysis, research and teaching. This is a challenging endeavour because, as Hamilton (2017) points out, the changes that are required are so radical that they constitute what philosopher of science Thomas Kuhn refers to as a 'paradigm shift'.

Thomas Kuhn's 'paradigm shifts': a brief overview

In his classic work *The Structure of Scientific Revolutions*, Kuhn defines 'paradigms' as 'universally recognized scientific achievements that for a time provide model problems and solutions to a community of practitioners' who are trained to engage in what Kuhn refers to as 'normal science' (Kuhn, 1996, p. x). Kuhn argues that in the course of doing this work, researchers sometimes encounter anomalies that cannot be explained by the existing paradigm and lead to crises that necessitate radical shifts, which he refers to as 'paradigm changes'. While Kuhn's work relates to enquiry in the natural sciences, Hamilton (2017) argues that the existing paradigms that social scientists work with cannot accommodate the anomaly that is represented by the advent of the Anthropocene, and that a paradigm shift is required in these discipline areas to deal with it. This shift in the social sciences is well overdue; the emergence of a new field of study, Earth System science, heralded such a paradigm shift in the natural sciences more than three decades ago.

Earth System science: a paradigm shift in the natural sciences

As ecosocialist Ian Angus explains, the study of the Earth as an integrated system originated in the 1980s when scientists realised that 'nuclear weapons, ozone-destroying chemicals, and greenhouse gases could radically remake the world: human activity was causing not just change but *global change*, with potentially disastrous consequences' (Angus, 2016, p. 30). The resultant scientific research in the emerging interdisciplinary field of Earth System science aimed to develop an understanding of how different components constituting the dynamic and complex natural world, including humans, interact (National Academy of Sciences, 1986). Over the years, Earth System scientists have systematically found evidence of widespread anthropogenic changes to the natural environment, including the disruption of major biogeochemical cycles such as the carbon, hydrological, nitrogen and phosphorous cycles that constitute the complex, dynamic biosphere of the planet (National Academy of Sciences, Engineering, and Medicine, 2017). Describing the dynamism of the Earth System, Galloway et al. (2014, p. 351) explain that Earth System components interact through several 'biogeochemical cycles' that function to conserve matter in the Earth System by shifting 'chemical elements among different parts of the Earth: from living to non-living, from atmosphere to land to sea, and from soils to plants'. Referring to the example of how reactive nitrogen has more than doubled in the biosphere since pre-industrial times, they furthermore draw attention to how '[g]lobal-scale alterations of biogeochemical cycles are occurring, from human activities both in the U.S. and elsewhere, with impacts and implications now and into the future' (*ibid.*).

So extensive are the geological and ecological impacts of human activities, and so profound their effects on the Earth's biosphere, that several Earth System scientists (for example, refer to Steffen et al., 2015) support atmospheric chemist Paul Crutzen's (2002) contention that they signal the commencement of a new geological epoch – the Anthropocene, or 'The Age of Humanity'. In contrast to the Holocene, the geological epoch which was characterised by the unusually stable climatic conditions and rich biodiversity that were conducive to the development and flourishing of human civilizations over the past approximately 11,500 years (Walker et al., 2009), the Anthropocene is characterised by a changing, unstable biosphere and, relatedly, the onset of what has been termed 'the sixth great extinction event' in our planet's history (McCallum, 2015). While the International Commission on Stratigraphy, the body whose task it is to formally determine and name geological time spans, is yet to decide on whether or not to formally acknowledge the Anthropocene as a new geological epoch, Angus (2016, p. 58) points out that even if it rejects the Anthropocene Working Group's recommendation to do so, this 'will not make the Anthropocene go away'.[3]

Apart from the larger controversy about whether or not humans have indeed become a geological force whose activities herald the Anthropocene, another important debate that has been preoccupying Earth System scientists and stratigraphers involves determining when the Anthropocene started, with suggestions ranging from Neolithic times to the Industrial Revolution and beyond (Oldfield et al., 2014). The starting point of the Anthropocene is very significant, as it can point to the root causes of the global changes, and identifying the cause of a problem is crucial if one is to address it effectively. Steffen et al. (2015) point out that while it is true that human activities affected *local* ecosystems prior to the 1940s, the changes their activities made to these ecosystems were neither extensive nor profound enough to have environmental effects on a planetary scale. It was only after the 'phenomenal growth of the global socio-economic system' in the 1950s, a phenomenon that Earth System scientists refer to as the 'Great Acceleration', that planetary-scale changes to natural systems became apparent (Steffen et al., 2015, p. 93). This is evident when one compares graphs depicting global socio-economic trends such as growths in population, real GDP, foreign direct investment, primary energy use, fertilizer consumption, transportation and international tourism (see Figure 1.1) with graphs showing biospheric system trends such as

Socio-economic trends

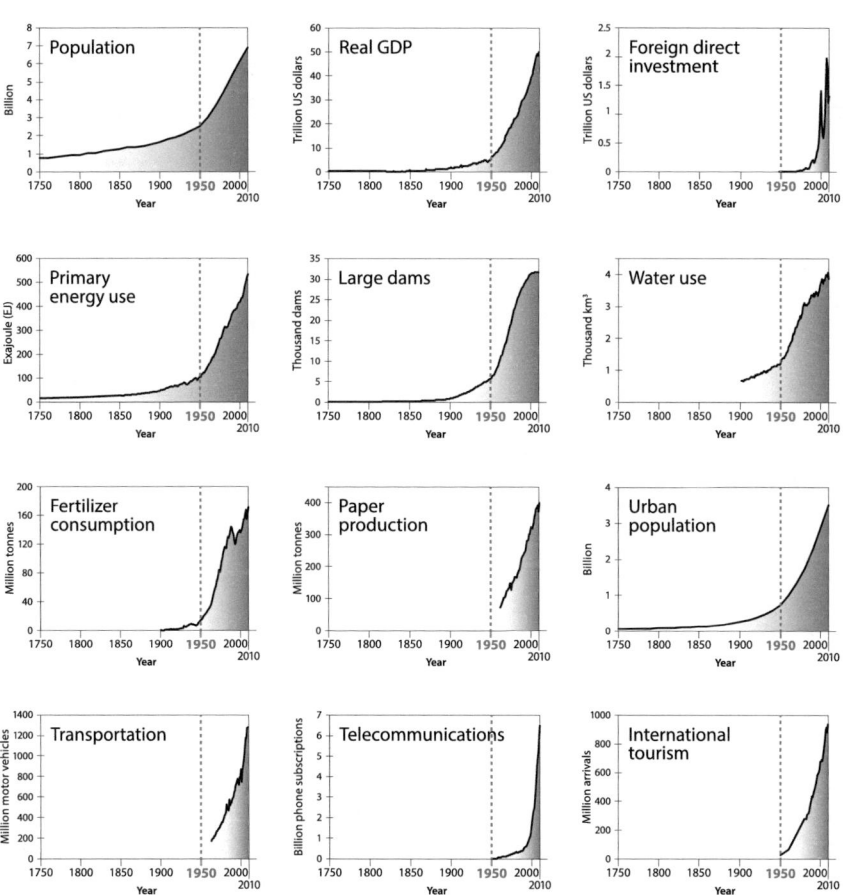

Figure 1.1 Socio-economic trends.[4]

Source: Steffen, W., et al. The trajectory of the Anthropocene: The Great Acceleration. *The Anthropocene Review* 2(1), pp. 81–98. Copyright © 2015 by the Authors. Reprinted by permission of SAGE Publications, Ltd.

Earth System trends

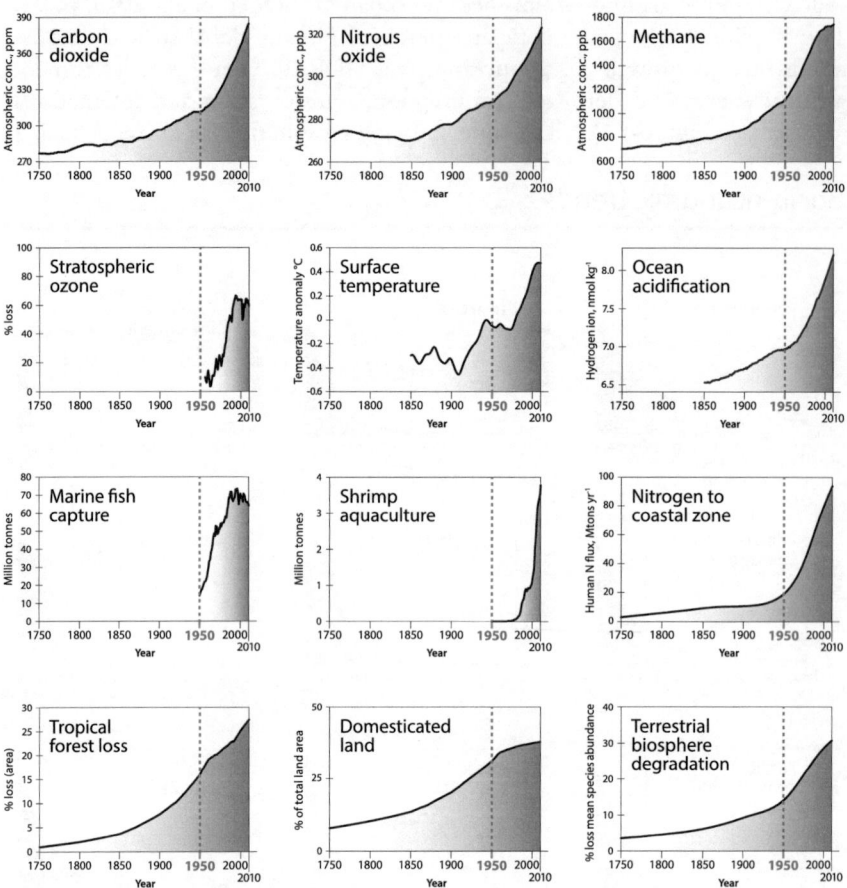

Figure 1.2 Earth System trends.

Source: Steffen, W., et al. The trajectory of the Anthropocene: The Great Acceleration. *The Anthropocene Review* 2(1), pp. 81–98. Copyright © 2015 by the Authors. Reprinted by permission of SAGE Publications, Ltd.

GHG concentrations, ocean acidification, surface warming and tropical forest loss over the period 1750–2010 (see Figure 1.2), which show concomitant sharp increases in all these indicators after 1950.

On the basis of this evidence, the preferred date that Will Steffen and his co-authors suggest for marking the onset of the Anthropocene is very precise.

> On Monday 16 July 1945, about the time that the Great Acceleration began, the first atomic bomb was detonated in the New Mexico desert.

Radioactive isotopes from this detonation were emitted to the atmosphere and spread worldwide entering the sedimentary record to provide a unique signal of the start of the Great Acceleration, a signal that is unequivocally attributable to human activities.

(Steffen et al., 2015, p. 93)

16 July 1945 would be a particularly apt symbolic marker for the onset of the Anthropocene as the energy released by atomic bombs is sometimes used in news reports and presentations to non-specialist audiences in order to help the public to better understand the incredible rate at which anthropogenic GHG emissions are disrupting the energy balance of our planet. For example, it is quite difficult for someone who is not used to working with large numbers to understand the implications of a recently published study of ocean warming. This study reports that:

Between 1987–2019, ocean warming was 450% greater than during the earlier time period [1955–1986] … the ocean heat anomaly (0–2000 m) in 2019 was 228 Zetta Joules (ZJ, $1 \text{ ZJ} = 10^{21}$ Joules) above the 1981–2010 average and 25 ZJ above 2018.

(Cheng et al., 2020, p. 137)

When Lijing Cheng, the lead author of the report, was interviewed about the study's results by a US Cable News Network (CNN) journalist, he used an atomic bomb explosion analogy to convey this information in a form that is easier for the non-specialist to understand: 'the amount of heat we have put in the world's oceans in the past 25 years equals to 3.6 billion Hiroshima atom-bomb explosions' (Kottasová, 2020). The article goes on to explain this further.

That's equivalent to dropping roughly four Hiroshima bombs into the oceans every second over the past quarter of a century. But because the warming is speeding up, the rate at which we are dropping these imaginary bombs is getting faster than ever. 'We are now at five or six Hiroshima bombs of heat each second,' said John Abraham, one of the authors of the study.

(Kottasová, 2020)

It is important to remember, however, that while radioactive isotopes in the sedimentary record serve as a potential *geological marker* for the beginning of the Anthropocene, it was the 'phenomenal growth of the global socio-economic system' after World War II (i.e., the global political economy established after World War II) that initiated the 'Great Acceleration trends' and the dynamic coupling of socio-economic and biophysical systems at a planetary scale (Steffen et al., 2015, pp. 93–94). While space limitations preclude a detailed discussion of the global political economy established in the aftermath of World War II, a brief overview of some of its key characteristics is necessary if we are to understand how human activity came to disrupt natural planetary processes as profoundly as Earth System science reveals.

Features of the post–World War II global political economy

According to some analyses, and as the graphs shown previously demonstrate, the period immediately following World War II witnessed 'the most successful phase of expansion in the history of capitalism' (Dunford, 2000, p. 151). As the primary victor of World War II, the United States collaborated with its allies to devise and implement a variety of measures with the aim of restoring conditions conducive to the expansion of capitalism, but particularly conducive to the interests of US-based capital (Budd, 2013).[5] These developments occurred in the wider context of the 'Cold War', which began in 1945 and ended (at least temporarily) with the 'collapse' of what was called the 'Eastern Bloc' in the early 1990s.

The Cold War period was characterised by a rivalry between an alliance of the advanced capitalist, industrially developed 'Western democracies' led by the US and its foremost ally, the United Kingdom, and the Eastern Bloc of 'communist' states led by the Union of Soviet Socialist Republics (USSR) or Soviet Union (Cox with Sinclair, 1996).[6] It had military, economic, political and ideological dimensions as those in power in the US- and Soviet-led blocs of nations attempted to spread and entrench their economic and political systems and, equally importantly, their ideologies throughout the globe (Cox with Sinclair, 1996; Gill and Law, 1988). The specific forms of social, political and economic arrangements that emerged in the advanced capitalist 'Western democracies' in the Cold War period were thus influenced by the existence of, and in competition with, the USSR and the Eastern Bloc (Budd, 2013).

As well as creating the intergovernmental military alliance the North Atlantic Treaty Organisation (NATO) in 1949 under the direction of the United States, leaders of the US-led alliance of advanced capitalist states sought to reinforce and strengthen their position in the international state system by incorporating and integrating more states into capitalist market relations.[7] The post-World War II period was thus characterised by the increasing integration of the advanced capitalist economies and other 'emerging economies' into the capitalist global economic system through a variety of international and supranational economic and political institutions (Cox, 1987; Gill and Law, 1988) whose role in managing global affairs has become increasingly important.

Post–World War II economic institutions: the Bretton Woods system

With the stated purposes of promoting 'economic growth' through 'free markets', the economic institutions established at the 1944 Bretton Woods Conference continue to facilitate the penetration of capitalist relations of

production into more regions of the global economy and also foster the continuing regional and global integration of capitalist economies (Gill, 1993).[8]
As noted previously, the US's military and economic dominance after World
War II gave it the leverage to shape the post-war world order that would
dictate the terms of economic relations amongst the Western capitalist states
in ways that favour the interests of US capital (Budd, 2013). An illustration of
the latter bias is evident in one of the key outcomes of the Bretton Woods
Conference: an International Monetary System (IMS) of fixed exchange rates
based on the US dollar which was, until 1971, valued with respect to gold
reserves (Ougaard, 2016). Acting as the world's primary reserve currency
since the mid-1940s, the US dollar is the currency most used by central banks
as a reserve asset as well as the currency most used when purchasing commodities and in the provision of international loans (Lucarelli, 2012). The
role of the US dollar as the world's primary reserve currency is favourable to
US government and business interests in a variety of ways, while it means
that US economic policies can cause financial (and hence social) crises in
other parts of the world (as discussed in the analysis of the global political
economy presented in Chapter 7).

An IMS based on the US dollar was accompanied by the creation of two
institutions charged with the task of overseeing the establishment and operation of a new international trade and credit system: the International
Monetary Fund (IMF) and the World Bank (Gill and Law, 1988). The establishment of these institutions in a way that favours US capital's interests is also
clearly evident; for example, voting in the World Bank and the IMF works
on a complex quota system that gives the US vote much greater weight than
that of other member states (IMF, 2016; World Bank, 2016).[9] Like the global
capitalist financial system established after World War II, the trade regime
established under the General Agreement on Tariffs and Trade (GATT), in
effect between 1947 and 1994, clearly favoured the economic interests of key
players in the advanced capitalist economies and discriminated against the
interests of the 'developing' economies by retaining protectionism in those
economic sectors that suit the advanced capitalist nations – that is, in sectors
such as agriculture and the textile and clothing industries, where developing
capitalist economies enjoy a 'comparative advantage' vis-à-vis advanced capitalist economies (McPhee, 1992).

David Ricardo's 'law of comparative advantage' is a cornerstone of liberal
economic theory and holds that in their trade relations with other nations,
'countries will tend to specialize in those commodities whose costs are com-
paratively lowest' (Gilpin, 1987, p. 173). In the 1980s, the neoclassical refor-
mulation of this theory maintained that 'a nation's comparative advantage is
determined by the relative abundance and most profitable combination of its
several factors of production, such as capital, labor, resources, management,
and technology' (Gilpin, 1987, p. 175). Absent from these accounts is any
notion of how asymmetrical power structures play any role in trade relations,
or how this theory's underlying assumptions of the immobility of the most

important factors of production (primarily capital, but also labour) are not valid in the current context of neoliberalising global capitalism so that practices justified on the basis of 'win-win' situations as a result of 'comparative advantage' become zero-sum games that favour the owners of capital, who enjoy an *absolute* advantage (Daly, 1993).

'Cold War' ideology: 'embedded liberalism' and liberal democracy

One important ideological obstacle to the establishment of a completely 'unregulated, free market' global economy facilitating unfettered capital accumulation was, however, presented by the existence of the Eastern Bloc. While this is seldom discussed in orthodox academic and mainstream literature, there is empirical evidence to justify claims of the superiority of the central planning model adopted by the Soviet Union in the post-war period. As Gill and Law (1988, p. 309) point out, 'growth rates for industrial production in [the] CMEA [Council for Mutual Economic Assistance] … consistently exceeded those for the EEC [European Economic Community] in the 1960s and 1970s (although these slowed considerably in the late 1970s and early 1980s).' There is also little or no unemployment in centrally planned economies (Gill and Law, 1988). In addition, relative to Western 'free market' economies, the post-World War II centrally planned economies were characterised by 'considerably less waste (there was no planned obsolescence, mass advertising, and unsalable surplus production, except where goods are defective)' (*ibid.*).

In summary, the Soviet central planning model for economic development, and social policies that provided secure jobs, free healthcare, education and housing, could not be ignored in the ideological and political struggle against 'communism'. For these reasons, the ideological struggle to demonstrate the attractiveness of capitalism during the Cold War played a major role in the domestic political economies of the industrially advanced capitalist states, whose governments were obliged to adopt welfare nationalist policies in order to alleviate the negative effects of private wealth accumulation and thus prevent social unrest and sympathy towards the opposing ideology of 'communism' (Dunford, 2000; Gilpin, 1987). This compromise was encapsulated by the term 'embedded liberalism', which was coined by John Ruggie (1982) and referred to the development of an international capital accumulation regime that achieved a compromise between the *laissez faire* 'free market' ideals of classical economics at the inter-state level and the need to provide a stable domestic political climate that facilitated continued capital accumulation at the national level. In the 1950s and 1960s, therefore, the social relations of production in the advanced capitalist economies were characterised by Fordism and Keynesianism, both being strategies designed to minimise unemployment rates and bolster consumer spending power, thereby providing governments with tools to smooth out the inevitable 'market failures' and ensuring social stability (Cox, 1987).

The need to present Western liberal (or 'representative') democratic systems as effective and superior mechanisms of government (vis-à-vis the competing Soviet model of state socialism) that were responsive to the needs and preferences of their populations was also important during this period. Liberal democratic theory contends that citizens have the power to influence policies adopted by their governments by voting in political parties that represent their interests. As Cox (1996, p. 532) notes, however, there is a formal separation between the political sphere (where citizens have 'equal rights') and the economic sphere (where private property has rights and citizens do not) in the classic meaning of liberal democracy.[10] At an international level, this separation between the political and the economic is also evident in the structure of the United Nations.

Post-World War II international institutions: the United Nations

While the International Monetary Fund (IMF) and the World Bank (now the World Bank Group) are the main economic institutions that were initially created to promote and manage the global spread of capitalism, a number of other international institutions were also established under the auspices of the United Nations (UN). The UN has six main organs and a variety of specialised agencies; however, an analysis of the UN Charter, its founding document, reveals that many UN institutions have no real power and seem to have been designed to perform the ideological role of legitimising the post-World War II global order rather than to facilitate any real 'power-sharing'.

The United Nations Security Council (UNSC) is the most powerful UN organ. It has five permanent members (the US, the UK, the Russian Federation, China and France), who have veto powers over decisions about war and peace. The Security Council's power is evident in many different ways, including in its ability to authorise economic sanctions against states that its members want to 'punish' for perceived infractions and its mandate to select the Secretary-General of the United Nations Secretariat, the administrative organ of the UN. The power of the UNSC is also evident in how its decisions are binding on all member states while the United Nations General Assembly (UNGA), which is ostensibly the most democratic organ of the UN because all member state representatives have the right to a vote (and all votes are weighted equally), only has the authority to make recommendations on a restricted range of non-budgetary issues. In addition, the UNGA is not permitted to make any recommendations on issues or disputes being considered by the UNSC. Like the UNGA, most of the UN's specialised agencies also have very little power.

Examples of UN specialised agencies are the World Health Organisation (WHO), the Food and Agricultural Organisation (FAO), the World Meteorological Organization (WMO) and the International Labour Organization (ILO). Many of these agencies were established to address specific issue areas or, as O'Brien and Williams (2010, p. 136) put it, 'to cope with the casualties of the global political economy'. UN institutions such as the IPCC and regulatory frameworks such as the United Nations Framework Convention on Climate Change (UNFCCC) have also been established to address the most significant 'casualty' of global capitalism discussed in this book: the Earth's biosphere. It is noteworthy that, in addition to being much more powerful than the other specialised agencies, economic institutions such as the IMF and the World Bank Group also take precedence over the IPCC and UNFCCC and present serious obstacles to taking effective action to mitigate AGW. This is particularly the case because the IPCC was only established in 1988 and the UNFCCC only came into force in 1994, well after the beginning of the 'neoliberal globalisation' project that signalled a renewed focus on expanding and intensifying capitalist relations of production that favour US capital after what some analysts describe as the 1970s 'structural crisis'.

Structural crisis of the post-World War II world order in the mid-1970s

As Cox (1987) points out, while the causes of the 1968–1975 crisis in the world order established after World War II under US leadership are open to debate, many political theorists agree that developments leading to the world recession that began in 1974 prompted a significant rearrangement of the operations and management of the post-World War II global political economy. Analysts' attempts to explain the causes of the post-war structural crisis of global capitalism vary in many ways, including in the issues they choose to focus on. For example, some theorists emphasise the falling rate of profit in the US, including the limitations placed on productivity efficiencies and economic growth by the Fordist production model (Rupert, 1997) and the relative increase in wages as a result of the power of trade unions. Representing the state-centric Realist international relations (IR) perspective, Gilpin (1987) identifies the crisis in terms of the end of the era of American economic dominance in the global economy as a result of several factors, including the increasing economic competitiveness of Japan, West Germany and the newly industrialising countries (NICs) and 'the first oil shock' presented by the quadrupling of world energy prices with the formation of the Organisation of Petroleum Exporting Countries (OPEC) in 1973–1974. O'Brien and Williams (2010) argue that while the imports of cheap manufactured goods threatened some jobs in the advanced capitalist states, the benefits of access to cheap imported consumer goods and the creation of employment opportunities in new economic sectors led to the willing acquiescence of workers to the trade liberalisation process that ensued as a result of this crisis. Cox (1987), however, argues that accounts emphasising worker complicity in the changes are incomplete because the restructuring of the global political economy also crucially involved governments adopting domestic policies, including the use of force, that greatly weakened the ability of workers to mount effective resistance to these changes. The role of the state in weakening domestic labour movements in the US, the UK and Australia during the 1980s has been extensively documented by many analysts, adding validity to Cox's critique.[11] Beyond the weakening of the labour movement, government leaders and policymakers of this 'advance wave' of what some analysts refer to as the 'neoliberalising capitalist project' were important agents, initiating and organising several other necessary policy changes to prepare the ground for the emergence of what is variously referred to as 'globalisation', 'neoliberalism' or 'neoliberal globalisation'.

While the balance of individual factors driving and facilitating the post-1970s changes to the global political economy are given different weight by different theorists, there is no doubt that technological developments, particularly in the information and communications technologies (ICT) sector, also played a central role in facilitating the further expansion of capitalism. However, some analysts (for example, see Amoore et al., 1997) argue that it is both simplistic and misleading to identify ICT innovation as the most important factor driving changes in the global political economy because the development and widespread adoption of these technologies were supported and promoted by government policies and funding (Jones, 1997). In addition, and more importantly, technological advances cannot explain the specific form that the current phase of the expansion of capitalism has taken: '*neoliberal* globalisation'. Although the concept of 'neoliberal globalisation' has been the subject of extensive disputes (as discussed further in Chapter 2), many analysts agree that the modifications to the global political economy that were initiated in the 1970s can be characterised in terms of several key features.

Overview of post-1970s modifications to the global political economy

The post-1970s modifications to the global political economy can be summarised as having been designed with the purpose of, and having the effect of, expanding both the operations and the intensity of global capitalism. These modifications included: the gradual dissolution of key elements of the Bretton Woods system, such as the gold standard; the increasing power of actors such as transnational corporations and financial agents such as hedge funds, credit ratings agencies and institutional investors (Helleiner and Pagliari, 2009); the deregulation of the finance sector and the creation of new financial instruments (Keaney, 2014; Lapavitsas, 2013); the widespread implementation of economic policies informed by a return to neoclassical economic orthodoxy that aim to 'deregulate' the global economy and privatise the public sector; and the increasing power extended to international economic institutions such as the World Bank, the International Monetary Fund and the World Trade Organization to implement and enforce these policies (Gill and Law, 1998).

These developments, in addition to the later collapse of the Soviet Union, signalled the beginning of an increasingly unstable period in global politics characterised by a rearrangement of national and global social relations of production and exchange, and concomitant important shifts in the functions and operations of nation states both at the domestic and global levels (Cox, 1987; Gill and Law, 1988). Given the current climate change crisis, however, the ecological damage caused by the expansion of global capitalism under the auspices of neoliberal globalisation processes is the most important feature of these developments.

Overview of the ecological implications of neoliberal globalisation processes

The expansion of capitalism, and its drive for ever-increasing 'economic growth', which is very narrowly defined in terms of gross domestic product (GDP) that does not account for what economists refer to as 'externalities' (such as GHG emissions and general environmental degradation), comes at a great cost as environmentally damaging production and transportation methods contaminate more regions of the world and unsustainable consumption patterns are encouraged both in the Global North and in the 'developing' economies of the Global South (Williams, 1996). Scientists warn that these business practices risk making the planet uninhabitable as the serious environmental degradation they cause threatens the viability of major ecosystems and of the entire biosphere within which humans are embedded (Steffen et al., 2011).

Environmental destruction takes many forms, all of which are important to address; however, scientists working in a variety of discipline areas agree that the most urgent current symptoms of the many interlinked environmental crises are the global warming, climate change and ocean acidification that result from historical and ongoing anthropogenic GHG emissions. These symptoms of a changing biosphere manifest as 'extreme weather' events such as prolonged droughts, powerful cyclones and unprecedentedly intense floods whose effects are exacerbated by sea-level rise and urbanisation. According to scientific bodies such as the World Meteorological Organization (2016), if we continue with 'business as usual', these phenomena and their consequences are likely to continue and intensify. The 2019/2020 Australian megafires demonstrate both the catastrophic ecological and socio-economic consequences of such 'extreme weather' events and the urgent need for a paradigm shift in how we think and act in the Anthropocene.

Lessons from the 2019/2020 Australian megafires: how to think in the Anthropocene

The unprecedented intensity and size of the 2019/2020 Australian bushfires attracted much publicity and raised widespread concern regarding what these 'megafires' signal about the effects of climate change (Sanderson and Fisher, 2020). Like other terminology used to describe the onset of 'extreme weather' events (for example, 'rain bombs' and 'bomb cyclones'), the term 'megafire' is relatively new, becoming part of the common discourse in the early 2000s when it was used to describe the 'massive blazes' in the United States 'that create more than usual destruction to life and property' (Read and Denniss, 2020). Large and intense enough to create their own weather systems, megafires generate 'winds that throw embers 30 kilometers or more ahead of the front and pyro-cumulus clouds [which] produce dry lightning that ignites new fires' (*ibid.*). There is a widespread concern that such 'megafires' may be becoming the 'new normal' in some regions of the world.

While it is difficult to attribute individual wildfires to the effects of AGW (Sanderson and Fisher, 2020), it is well documented that the climate change that results from AGW is increasing the length and frequency of heatwaves, and that higher average temperatures exacerbate drought conditions and result in longer bushfire seasons (Steffen et al., 2019). There is evidence that such conditions prevailed before the outbreak of these 'unprecedented' megafires. In its 2019 *Annual Climate Statement*, the Australian Bureau of Meteorology (2020) identifies 2019 as 'Australia's warmest year on record; marked by severe, protracted drought', and states that 'warming associated with anthropogenic climate change has seen Australian annual mean temperatures increase by over one degree since 1910'. While investigations regarding the extent to which climate change contributed to the 2019/2020 Australian megafires are still ongoing, scientists are concerned 'that these unprecedented fires may indicate that the more flammable future projected to eventuate under climate change has arrived earlier than anticipated' (Boer, de Dios and Bradstock, 2020). The devastating effects of this 'more flammable future' on ecosystems and people are particularly concerning.

The ecological, social and economic implications of a 'more flammable future'

Officials estimate that more than 20% of Australia's forests burned in the 2019/2020 bushfire season, 'a proportion [that] scientists believe is unprecedented globally' (Boer, de Dios and Bradstock, 2020). In addition to killing 33 people, the 2019/2020 megafires have burned more than 12.6 million hectares to date (Werner and Lyons, 2020) and caused economic damage that is estimated to amount to nearly AUD $100 billion (Read and Denniss, 2020). These fires also caused widespread ecological damage, emitting 434 million tonnes of carbon dioxide (CO_2) into the atmosphere and killing over a billion animals and over 200 billion invertebrates (Werner and Lyons, 2020). In addition to these immediate catastrophic effects, urban researcher Anthony Richardson (2020) outlines how the 2019/2020 bushfires 'highlighted the immense vulnerability of the systems which make our contemporary lives possible'.

Contextualising 'natural disasters': complex socio-ecological systems

Richardson (2020) points out that during the Australian megafires, road closures and power cuts led to fuel and food shortages as well as to disruptions to mobile phone services, automatic teller machines and electronic fund transfer at point of sale (EFTPOS) services. He argues that these developments demonstrate the need to understand the cascading 'political, technological, economic or social' effects that a change in one part of a complex system, such as the change to a landscape that is caused by a bushfire, can initiate. While Richardson's analysis focuses on the unfolding of a disaster at a local level, Homer-Dixon et al. (2015) explore 'how a catastrophic social-ecological crisis of global scope might occur'. They argue that such a crisis could begin as localised, smaller crises that spread, combine and expand globally. Understanding and trying to pre-empt such a global social-ecological crisis requires researchers to 'closely investigate the complex, dense, and evolving causal links among humanity's energy, food, water, climate, and financial systems' (*ibid.*). However, researchers wishing to adopt such a holistic approach face formidable obstacles, including the limitations they encounter by the disciplinary confines of their fields of study.

Specialisation versus complexity in the Anthropocene

In contrast to the complexity of the ways in which social and natural systems interact in the real world, most academics working within the 'Western' intellectual tradition are trained, and required, to study and understand aspects of this totality by specialising in different discipline areas. This is because, unlike the holistic understanding that some indigenous peoples still have of the interrelationships between humans and the natural systems in which they live (Ludwig and Poliseli, 2018), the dominant culture that developed hand-in-hand with capitalism separates the 'natural' from the 'social' so that an artificially created strict separation between the 'natural sciences' and the 'social sciences' is generally taken for granted in educational institutions (van der Pijl, 2009; see also van der Pijl, 2012) as well as in everyday life.

Both the natural sciences and the social sciences are furthermore divided into distinct discipline areas, each specialising in its own field of study which is delimited by strict boundaries regarding appropriate content for that field. An offshoot of political science, international relations (IR) is a case in point of the 'disciplinary boundary' issues that arise when trying to delimit areas of investigation (for example, refer to Albert and Buzan, 2017). These artificial boundaries are extremely problematic as they yield reductionist analyses that are too simplistic to account for what happens in the real world (Veltmeyer, 2019). In the social sciences, for instance, political science and economics exist as distinct discipline areas, despite the fact that in the real world the issues they study are so tightly interconnected that it is difficult to truly understand them without using an integrated 'political economy' approach (*ibid.*; see also van der Pijl, 2009). Realism, which is the dominant IR perspective, is similarly unable to account for the complexity that exists in the real world as it focuses on relations between nation states in the 'state system' while ignoring the economic and social forces that constitute the states themselves. In contrast, the scientific understanding of the unprecedented ways in which current economic activities damage the environment presents an urgent challenge to researchers to develop *holistic* understandings of the effects of the social systems embedded within our complex, dynamic Earth System. As noted previously, such holistic understandings will entail adopting a revolutionary paradigm shift in the Western intellectual tradition.

Holocene and Anthropocene social sciences: the need for a paradigm shift

Elaborating on the implications of Earth System scientists' observation that 'We are now living in a no-analogue world' (Steffen et al., 2015, p. 94), Hamilton emphasises the radical implications of the Anthropocene.

> It is of utmost importance to understand that the 'Anthropocene' is not a term coined to describe the continued spread of human impacts on the landscape or further modifications to ecosystems; it is instead a term describing a *rupture* in the functioning of the Earth System as a whole, so much so that the Earth has now entered a new geological epoch.
>
> (Hamilton, 2017, pp. 9–10, emphasis in original)

As noted previously, Hamilton (2017) argues that these developments call for a similarly radical and revolutionary paradigm shift in all discipline areas within the academy. Scientific papers also emphasise the need to integrate knowledge from the social sciences in Earth System science while acknowledging the difficulty of doing so, as is evident in this extract from a 2017 US National Academy of Sciences report.

> Although the need to integrate social science research within the [United States Global Change Research] Program [USGCRP] was recommended early in its existence, there were and are obstacles to this effort. Most universities are organized primarily around disciplines, although some have also developed innovative structures to encourage interdisciplinary research and education. Most of the USGCRP agencies, as well as the national laboratories, have very few social scientists on staff, making integration difficult.
>
> (National Academy of Sciences, Engineering, and Medicine, 2017, p. 24)

An increasing number of academics working in humanities, arts and social science disciplines are now trying to address the challenges of the Anthropocene as these relate to their fields of enquiry, and some of them also argue that new approaches are required to understand and address the complexity of the interactions between human and natural systems. There have also been attempts to include social scientists in Earth System science programme teams such as those associated with USGCRP agencies, but the ontological and epistemological obstacles they face are formidable and are discussed in more detail in the next chapter.

Notes

1 The IPCC (2018b, p. 10) explains that 'pre-industrial' refers to the period from 1850 to 1900.
2 The IPCC (2007, p. 875) defines an 'extreme weather event' as

> [a]n event that is rare within its statistical reference distribution at a particular place. Definitions of 'rare' vary, but an extreme weather event would normally be as rare as or rarer than the 10th or 90th percentile. By definition, the characteristics of what is called 'extreme weather' may vary from place to place. Extreme weather events may typically include floods *and* droughts.

Scientists have been warning for many years that what mainstream media report as 'extreme weather events' is becoming the 'new normal' (IPCC, 2012). Defining the 'new normal' as 'the point in time when at least half the following twenty years are warmer than 2015's record breaking global temperatures', Lewis (2016) describes the findings of the research team she worked with as 'straightforward'. In Australia, which is 'the canary in the coal mine', '2015's record-breaking temperatures will be the new normal between 2020 and 2030 according to most of the climate models we analysed' (*ibid.*).
3 The concept of the 'Anthropocene' is the subject of several controversies, including important critiques of the political implications of the term itself (for example, refer to Malm and Hornborg, 2014). Swyngedouw and Ernstson (2018, p. 3), for instance, argue that the Anthropocene is a:

> 'deeply depoliticizing notion' that could prove extremely dangerous by opening 'the spectre (albeit by no means necessarily or intentionally…) for a deepening of a hyper-accelerationist eco-modernist vision in which big science, geo-engineering, and big capital can gesture to save both earth and earthlings.
>
> (*ibid.*, p. 3)

Other theorists argue that because the cause of the planetary changes are specifically capitalist relations of production rather than 'humans' in general, it would be more appropriate to refer to the new geological epoch as the 'Capitalocene' (Moore, 2016). While there are significant disagreements amongst both theorists and activists about whether or not the word 'Anthropocene' should be replaced by the word 'Capitalocene', ecosocialists generally refer to the Anthropocene (as does most of the academic literature) and I also follow that convention in this book.
4 The material reprinted in Figures 1.1 and 1.2 is the exclusive property of SAGE Publishing and is protected by copyright and other intellectual property laws.

5 Definitions of terms such as 'capitalism' and 'capital' are discussed in more detail in Chapter 2, which provides overviews of the theoretical perspectives used in this book as well as meanings of related key concepts.

6 The appropriateness of the use of the term 'communist' to describe the politico-economic systems in existence in the Eastern Bloc is contested; nevertheless, traditional IR accounts refer to that system as 'communist'.

7 It is not widely known that the1955 integration of West Germany into NATO was the event that prompted the creation of the Soviet-led Warsaw Pact (Kaplan, 1969).

8 'Free markets' are, paradoxically, always enforced by strict global trade rules backed up by sanctions for countries that do not comply. These rules greatly favour the groups who have the power to develop and implement them.

9 In 2015, the quota of the US vote in the IMF and the World Bank was more than double that of Japan, the member state with the second-highest voting quota in these institutions. An examination of the voting quotas listed on the relevant IMF and World Bank webpages demonstrates that 'developing economy' member states are given very little say in the decisions made in these organisations.

10 Cox argues that this distinction lost some coherence in the Cold War era, during which 'people used their political rights to limit and to channel the rights of property – to correct the inequities of the market' (Cox with Sinclair, 1996, p. 532).

11 For example, refer to Cahill (2014) for a discussion of the Australian case and to Cox (1987), Harvey (2005) and Klein (2007) for analyses of the US and British cases.

References

Albert, M. and Buzan, B. (2017). On the Subject Matter of International Relations. *Review of International Studies*, 43(5), 898–917.

Amoore, L., Dodgson, R., Gills, B.K., Langley, P., Marshall, D. and Watson, I. (1997). Overturning 'Globalisation': Resisting the Teleological, Reclaiming the 'Political'. *New Political Economy*, 2(1), 179–195.

Angus, I. (2016). *Facing the Anthropocene: Fossil Capitalism and the Crisis of the Earth System*. New York: Monthly Review Press.

Australian Bureau of Meteorology (2020). Annual Climate Statement 2019. 9 January. Available at www.bom.gov.au/climate/current/annual/aus/.

Boer, M.M., de Dios, V.R. and Bradstock, R.A. (2020). Unprecedented Burn Area of Australian Mega Forest Fires. *Nature Climate Change*, 10(3), 171–172.

Budd, A. (2013). *Class, States and International Relations: A Critical Appraisal of Robert Cox and Neo-Gramscian Theory*. London: Routledge.

Cahill, D. (2014). *The End of Laissez-Faire? On the Durability of Embedded Neo-liberalism*. Cheltenham: Edward Elgar Publishing.

Cheng, L., Abraham, J., Zhu, J., Trenberth, K.E., Fasullo, J., Boyer, T., Locarnini, R., Zhang, B., Yu, F., Wan, L., Chen, X., Song, X., Liu, Y. and Mann, M.E. (2020).

Record-Setting Ocean Warmth Continued in 2019. *Advances in Atmospheric Sciences*, 37(2), 137–142.

Cox, R.W. (1981). Social Forces, States, and World Orders: Beyond International Relations Theory. *Millennium: Journal of International Studies*, 10(2), 126–155.

Cox, R.W. (1983). Gramsci, Hegemony and International Relations: An Essay in Method. *Millennium: Journal of International Studies*, 12(2), 162–175.

Cox, R.W. (1987). *Production, Power and World Order: Social Forces in the Making of History*. New York: Columbia University Press.

Cox, R.W. (1996). Globalization, Multilateralism, and Democracy (1992). In Cox, R.W. with Sinclair, T.J., *Approaches to World Order*. Cambridge: Cambridge University Press, 524–536.

Cox R.W. with Schechter, M.G. (2002). *The Political Economy of a Plural World: Critical Reflections on Power, Morals and Civilization*. New York: Routledge.

Cox, R.W. with Sinclair, T.J. (1996). *Approaches to World Order*. Cambridge: Cambridge University Press.

Crutzen, P.J. (2002). The Anthropocene: Geology of Mankind. *Nature*, 415(23), 23.

Daly, H.E. (1993). The Perils of Free Trade: Economists Routinely Ignore its Hidden Costs to the Environment and the Community. *Scientific American*, 269(5), 24–29.

Dunford, M. (2000). Globalization and Theories of Regulation. In R. Palan (ed.), *Global Political Economy: Contemporary Theories*. London: Routledge, 143–167.

Galloway, J.N., Schlesinger, W.H., Clark, C.M., Grimm, N.B., Jackson, R.B., Law, B.E., Thornton, P.E., Townsend, A.R. and Martin, R. (2014). Biogeochemical Cycles. In Melillo, J.M., Richmond, T.C. and Yohe, G.W. (eds), *Climate Change Impacts in the United States: The Third National Climate Assessment*. US Global Change Research Program. Washington, DC: US Government Printing Office, 350–368.

Gill, S. (ed.) (1993). *Gramsci, Historical Materialism and International Relations*. Cambridge: Cambridge University Press.

Gill, S. and Law, D. (1988). *The Global Political Economy: Perspectives, Problems, and Policies*. Baltimore, MD: The Johns Hopkins University Press.

Gilpin, R. (1987). *The Political Economy of International Relations*. Princeton, NJ: Princeton University Press.

Hamilton, C. (2017). *Defiant Earth: The Fate of Humans in the Anthropocene*. Crows Nest, NSW: Allen & Unwin.

Harvey, D. (2005). *The New Imperialism*. New York: Oxford University Press.

Helleiner, E. and Pagliari, S, (2009). Towards a New Bretton Woods? The First G20 Leaders Summit and the Regulation of Global Finance. *New Political Economy*, 14(2), 275–287.

Homer-Dixon, T., Walker, B., Biggs, R., Crépin, A-S., Folke, C., Lambin, E.F., Peterson, G.D., Rockström, J., Scheffer, M., Steffen, W. and Troell, M. (2015). Synchronous Failure: The Emerging Causal Architecture of Global Crisis. *Ecology and Society*, 20(3), 6.

Intergovernmental Panel on Climate Change (IPCC) (2007). *Impacts, Adaptation and Vulnerability. Contribution of Working Group II to the Fourth Assessment Report of the Intergovernmental Panel on Climate Change*. Parry, M.L., Canziani, O.F., Palutikof, J.P., van der Linden, P.J. and Hanson, C.E. (eds). Cambridge: Cambridge University Press.

Intergovernmental Panel on Climate Change (IPCC) (2012). *Managing the Risks of Extreme Events and Disasters to Advance Climate Change Adaptation, A Special Report of*

Working Groups I and II of the Intergovernmental Panel on Climate Change. Field, C.B., Barros, V., Stocker, T.F., Qin, D., Dokken, D.J., Ebi, K.L., Mastrandrea, M.D., March, K.J., Plattner, G.-K., Allen, S.K., Tignor, M. and Midgley, P.M. (eds). Cambridge and New York: Cambridge University Press.

Intergovernmental Panel on Climate Change (IPCC) (2018a). Summary for Policymakers. In *Global Warming of 1.5 °C. An IPCC Special Report on the Impacts of Global Warming of 1.5 °C above Pre-Industrial Levels and Related Global Greenhouse Gas Emission Pathways, in the Context of Strengthening the Global Response to the Threat of Climate Change, Sustainable Development, and Efforts to Eradicate Poverty*. Masson-Delmotte, V., Zhai, P., Pörtner, H.-O., Roberts, D., Skea, J., Shukla, P.R., Pirani, A., Moufouma-Okia, W., Péan, C., Pidcock, R., Connors, S., Matthews, J.B.R., Chen, Y., Zhou, X., Gomis, M.I., Lonnoy, E., Maycock, T., Tignor, M. and Waterfield, T. (eds). Geneva: World Meteorological Organization.

Intergovernmental Panel on Climate Change (IPCC) (2018b). *Understanding the IPCC Special Report on 1.5 °C*. Geneva: World Meteorological Organization.

International Monetary Fund (IMF) (2016). International Monetary Fund homepage. Available at www.imf.org/external/index.htm.

Jones, R.J.B. (1997). Globalisation versus Community. *New Political Economy*, 2(1), 39–51.

Kaplan, L.A. (1969). The United States and the Origins of NATO 1946–1949. *The Review of Politics*, 31(2), 210–222.

Keaney, M. (2014). Financialization and Social Structures of Accumulation Theory. *World Review of Political Economy*, 5(1), 45–77.

Klein, N. (2007). *The Shock Doctrine: The Rise of Disaster Capitalism*. London: Penguin, London.

Kottasová, I. (2020). Oceans are Warming at the Same Rate as if Five Hiroshima Bombs were Dropped in Every Second. CNN, 13 January. Available at https://edition.cnn.com/2020/01/13/world/climate-change-oceans-heat-intl/index.html.

Kuhn, T.S. (1996). *The Structure of Scientific Revolutions* (3rd edn). Chicago, IL: University of Chicago Press.

Lapavitsas, C. (2013). The Financialisation of Capitalism: 'Profiting without Producing'. *City*, 17(6), 792–805.

Lewis, S. (2016). 2015's Record-Breaking Temperatures will be Normal by 2030 – it's Time to Adapt. The Conversation, 7 November. Available at https://theconversation.com/2015s-record-breaking-temperatures-will-be-normal-by-2030-its-time-to-adapt-68224.

Lucarelli, B. (2012). Financialization and Global Imbalances: Prelude to Crisis. *Review of Radical Political Economics*, 44(4), 429–447.

Ludwig, D. and Poliseli, L. (2018). Relating Traditional and Academic Ecological Knowledge: Mechanistic and Holistic Epistemologies across Cultures. *Biology and Philosophy*, 33(5–6), 43.

Malm, A. and Hornborg, A. (2014). The Geology of Mankind? A Critique of the Anthropocene Narrative. *The Anthropocene Review*, 1(1), 62–69.

McCallum, M.L. (2015). Vertebrate Biodiversity Losses Point to a Sixth Mass Extinction. *Biodiversity and Conservation*, 24(10), 2497–2519.

McPhee, J.S. (1992). Agriculture and Textiles: The Fare and Fabric of Current GATT Negotiations. *Indiana International and Comparative Law Review*, 3(1), 155–175.

Moore, J.W. (ed.) (2016). *Anthropocene or Capitalocene? Nature, History, and the Crisis of Capitalism*. Oakland, CA: PM Press.

National Academy of Sciences, Earth System Sciences Committee of the National Aeronautics and Space Administration Advisory Council (1986). *Earth System Science Overview: A Program for Global Change*. Washington, DC: The National Academies Press.

National Academy of Sciences, Engineering, and Medicine (2017). *Accomplishments of the US Global Change Research Program*. Washington, DC: The National Academies Press.

O'Brien, R. and Williams, M. (2010). *Global Political Economy: Evolution and Dynamics* (3rd edn). New York: Palgrave Macmillan.

Oldfield, F., Barnosky, A.D., Dearing, J., Fischer-Kowalski, M., McNeill, J., Steffen, W. and Zalasiewicz, J. (2014). The Anthropocene Review: Its Significance, Implications and the Rationale for a New Transdisciplinary Journal. *The Anthropocene Review*, 1(1), 3–7.

Ougaard, M. (2016). The Reconfiguration of the Transnational Power Bloc in the Crisis. *European Journal of International Relations*, 22(2), 459–582.

Read, P. and Denniss, R. (2020). With Costs Approaching $100 Billion, the Fires are Australia's Costliest Natural Disaster. The Conversation, 17 January. Available at https://theconversation.com/with-costs-approaching-100-billion-the-fires-are-australias-costliest-natural-disaster-129433.

Richardson, A. (2020). No Food, No Fuel, No Phones: Bushfires Showed We're Only Ever One Step from System Collapse. The Conversation, 7 February. Available at https://theconversation.com/no-food-no-fuel-no-phones-bushfires-showed-were-only-ever-one-step-from-system-collapse-130600.

Ruggie, J.G. (1982). International Regimes, Transactions and Change: Embedded Liberalism in the Post-War Economic Order. *International Organization*, 36(2), 397–415.

Rupert, M. (1997). Globalisation and American Common Sense: Struggling to Make Sense of a Post-Hegemonic World. *New Political Economy*, 2(1), 105–116.

Sanderson, B.M. and Fisher, R.A. (2020). A Fiery Wake-Up Call for Climate Science. *Nature Climate Change*, 10, 175–177.

Sinclair, T.J. (2016). Robert W. Cox's Method of Historical Structures Redux. *Globalizations*, 13(5), 510–519.

Steffen, W., Broadgate, W., Deutsch, L., Gaffney, O. and Ludwig, C. (2015). The Trajectory of the Anthropocene: The Great Acceleration. *The Anthropocene Review*, 2(1), 81–98.

Steffen, W., Hughes, L., Mullins, G., Bambrick, H., Dean, A. and Rice, M. (2019). *Dangerous Summer: Escalating Bushfire, Heat and Drought Risk*. Climate Council of Australia Ltd. Available at www.climatecouncil.org.au/resources/dangerous-summer-escalating-bushfire-heat-drought-risk/.

Swyngedouw, E. and Ernstson, H. (2018). Anthropo-obScene: Immuno-Biopolitics and Depoliticizing Ontologies in the Anthropocene. *Theory, Culture & Society*, 35(6), 3–30.

United Nations Environment Program (UNEP) (2019). *Emissions Gap Report 2019*. Nairobi: UNEP.

van der Pijl, K. (2009). A Survey of Global Political Economy. Available at www.academia.edu/31081267/A_SURVEY_OF_GLOBAL_POLITICAL_ECONOMY.

van der Pijl, K. (2012). The Limits of Discipline, or How to Make Sense of the State of the World. *Global Change, Peace and Security*, 24(1), 25–30.

Veltmeyer, H. (2019). Capitalism, Development, Imperialism, Globalization: A Tale of Four Concepts. *Globalizations*. DOI:10.1080/14747731.2019.1699706.

Walker, M., Johnsen, S., Rasmussen, S.O., Popp, T., Steffensen, J-P., Gibbard, P., Hoek, W., Lowe, J., Andrews, J., Björck, S., Cwyner, L.C., Hughen, K., Kershaw, P., Kromer, B., Litt, T., Lowe, D.J., Nakagawa, T., Newnham, R. and Schwander, J. (2009). Formal Definition and Dating of the GSSP (Global Stratotype Section and Point) for the Base of the Holocene using the Greenland NGRIP Ice Core, and Selected Auxiliary Records. *Journal of Quaternary Science*, 14(1), 3–17.

Werner, J. and Lyons, S. (2020). The Size of Australia's Bushfire Crisis Captured in Five Big Numbers. Australian Broadcasting Corporation (ABC), 5 March. Available at www.abc.net.au/news/science/2020-03-05/bushfire-crisis-five-big-numbers/12007716.

Williams, M. (1996). International Political Economy and Global Environmental Change. In J. Volger and M.F. Imber (eds), *The Environment and International Relations*. London: Routledge, 41–58.

World Bank (2016). Voting Powers. Available at www.worldbank.org/en/about/leadership/votingpowers.

World Meteorological Organization (WMO) (2016). (Un)Natural Disasters: Communicating Linkages between Extreme Events and Climate Change. *World Meteorological Organization Bulletin*, 65(20).

2 Theoretical perspectives for the Anthropocene

An ecological neo-Gramscian Method of Historical Structures

Introduction

While this is not always acknowledged, all inquiry in the social sciences is conducted through the lens of specific theoretical perspectives that are based on analysts' beliefs about the nature of social reality (ontology) and how best to study and understand this reality (epistemology). Ontological and epistemological beliefs are interrelated and also influence choices of appropriate research methodologies. These choices are important because theory is not disconnected from the real world; theory informs action. This chapter provides an overview of the ontological and epistemological choices informing the analysis presented in the rest of the book, and introduces a range of important concepts as well as the theoretical perspective used in the analysis presented in subsequent chapters.

Despite the growing awareness of the need for holistic approaches alluded to in the previous chapter, current research and policy approaches treat 'natural disasters' and other crises as discrete problems that governments must address and solve when they arise. However, treating 'economy', 'environment', 'politics' and 'society' as separate spheres of inquiry not only limits our understanding of the complex interactions between natural and social systems, but it also makes it more difficult to identify the underlying causes of the problems being investigated. In contrast to such disconnected approaches, the Marxist analytical perspective has always been a holistic study of the 'political economy' of capitalism. Marx and Engels analysed capitalism in terms of the antagonistic social relations that it engenders, thereby seamlessly integrating politics and economics in their analysis of the capitalist mode of production. In addition, the holistic nature of the analytical perspective developed by Karl Marx has always included references to the environmental damage caused by the capitalist mode of production and, as discussed in more detail in the chapter on ecosocialist theory (Chapter 3), ecosocialists such as John Bellamy Foster and Paul Burkett demonstrate its inherent suitability for analysing and understanding ecological issues like the current anthropogenic climate change crisis. In this chapter, the overview of Marx's general analysis of capitalism introduces many of the concepts ecosocialists use in their analysis.

Given the controversies surrounding the meanings of Marxist concepts, however, it is useful to begin by briefly reviewing the notion of 'essentially contested concepts'. In addition to explaining why the meanings of many Marxist concepts are disputed, the discussion of 'essentially contested concepts' also aims to highlight the ideological component of all analyses in the social sciences and to thereby acknowledge the contested nature of many of the key concepts used in this book.

Essentially contested concepts and their ideological role

It is widely acknowledged that the meanings of key social science concepts such as 'democracy', 'social justice' and 'freedom' are 'essentially contested'. Gallie (1956, p. 169) defines 'essentially contested concepts' as those concepts that 'inevitably involve endless disputes about their proper uses on the part of their users', and he proposes several criteria whereby such concepts can be identified. Essentially contested concepts are: 'appraisive' (evaluative, thereby implicitly embodying normative positions); internally complex (comprising of combinations of several features that can be differently ranked in order of importance); initially diversely describable (so that alternative explanations exist from the outset because of their internal complexity); 'open' (in the sense that their meaning can change if new situations develop); and recognised to have multiple interpretations that are used both aggressively and defensively (Gallie, 1956).

Because of their complexity and the multidimensional phenomena they refer to, and particularly because of the normative *values* they implicitly embody, the meanings of essentially contested concepts are ambiguous and 'persistently vague' so that 'different persons or parties adhere to different views of the[ir] correct use' (Gallie, 1956, p. 172). For example, there are entire books devoted to defining the meaning of the concept of 'democracy' (refer to Cunningham, 2002 and Roper, 2013 for two such examples). This is because 'democracy' is a very complex concept encompassing many phenomena (such as 'rights') that are also open to a variety of interpretations. Despite this complexity, theorists who refer to 'democracy' very seldom define what they mean, and their meaning may not align with the meaning that their readers (or other theorists) assign to this concept. While the resulting intellectual confusion is sometimes accidental (arising either from commentators themselves using such concepts inconsistently or from an assumption that everyone else understands and agrees with their own definitions), at other times different understandings of the meanings of key concepts arise as a result of 'conceptual *contestation*' (Collier, Hidalgo and Maciuceanu, 2006, p. 212; emphasis in original). While interpretive contestations are often ideologically driven, they are sometimes motivated by the desire to refine understandings of the complex concepts under consideration. The latter motivation informs many debates amongst Marxists about the meanings and contemporary relevance of 'internally complex' concepts such as 'class'.[1]

Ideological implications of different interpretations of 'essentially contested concepts'

Concepts such as 'the Anthropocene', 'globalisation' and 'neoliberalism' (as well as many other key concepts used in this book) are, by Gallie's criteria, clearly essentially contested. What is important to understand, however, is that debates about the meaning of such key concepts are not only intellectually interesting but have important practical implications as they either support the status quo or they present critical challenges to it. Another way of thinking about 'essentially contested concepts' is as 'boundary concepts' (de Lucia, 2014) whose meaning is of vital importance because it informs possible actions that influence future trajectories. As discussed in more detail in Chapters 4, 5 and 6, for example, different definitions of contested concepts such as 'climate justice' and 'sustainability' have very real implications for people, for the integrity of ecosystems and for the fates of non-human life-forms. For the purposes of this book's arguments, the most important 'essentially contested' concepts that need to be defined at the outset are 'capitalism' and related Marxist concepts. These concepts are particularly complex, and the following rudimentary summary draws extensively on Bottomore (1991) and Fine and Saad-Filho (2016) to provide a brief (and necessarily simplistic) introduction to some aspects of Marx and Engels' analysis of capitalism.

A Marxist overview of the origins and key features of capitalism

One Marxist definition of capitalism is that it is 'a particular form of class society … in which a surplus product is extracted from direct producers' (Bottomore, 1991, p. 253). While Marx was not unaware of additional class categories, the two major classes he identifies in capitalist societies are the bourgeoisie (or capitalist class), which owns the means of production, and the proletariat (or working class). Having been deprived of access to their own means of production (and therefore to an independent way of securing their own subsistence), workers have no option but to sell their labour power in the 'market place' in order to survive. It is interesting to note that Marx himself 'does not use capitalism as a noun either in the *Communist Manifesto* or in *Capital I*' (Bottomore, 1991, p. 72); however, a brief overview of the way he refers to the concept 'capital' can usefully highlight some of the key features of the capitalist mode of production.

Capital as an antagonistic social relation between the capitalist class and the working class

Contrary to the common, trans-historical understanding that the term 'capital' refers to 'assets' that individuals have used in all societies (and will use in all future societies) to generate income, Marx identifies capital as a defining feature of *capitalist* societies – or, more accurately, of the capitalist mode of production (Bottomore, 1991). Moreover, Marx does not consider capital as a 'thing' (*ibid.*). To Marx, capital is, in essence, a *social relation* between classes – an *antagonistic* social relation that involves a dynamic class struggle between capitalists (who try to increase their profits by maximising the rate of exploitation of their workers) and workers (who try to minimise their own exploitation). In addition to being a social relation, Marx also describes capital as 'self-expanding value'. This feature of capital requires further explanation, as it is central to the argument that capitalism cannot solve the urgent ecological crises – including the climate change crisis – that we find ourselves in. To understand why capital is 'self-expanding value', however, it is necessary to have at least a rudimentary understanding of Marx's concept of 'value'.

Marx's labour theory of value

Marx identifies the production of commodities as a central feature of the capitalist mode of production, and defines commodities as 'use values produced by labour for exchange' (Fine and Saad-Filho, 2016, p. 17; emphasis in original). A 'mode of production' refers to the way in which a society is organised to produce the items necessary for the survival and reproduction of both its members and the mode of production itself, and all modes of production incorporate a social division of labour that facilitates the production of necessary items (both material and immaterial) that have a 'use value' (*ibid.*, p. 14). Marx distinguishes between different modes of production with reference to their 'forces of production' (which include the means of production and labour power) and the dominant 'relations of production' (which arise from who owns the means of production) (Bottomore, 1991). The fundamental relations of production in the capitalist mode of production are, as noted previously, antagonistic, and result from the fact that capitalists own the principal means of production, which is 'capital in its various forms' (Bottomore, 1991, p. 71), while workers own only their own labour power and are compelled to sell it as a commodity in exchange for a wage (the distribution of which is also controlled by capitalists) if they are to survive. At this point, it is important to note that while the use value of labour power is its ability to create other use values in *all* modes of production, it has the unique property of being the only force of production that creates *value* in capitalist societies (Fine and Saad-Filho, 2016, p. 21). This insight constitutes the basis of Marx's 'labour theory of value'.

In developing his labour theory of value, Marx observes that exchanging one use value (for example, a shirt) for another use value of a different sort (for example, a shoe) entails the commodities being homogeneous in some way, and argues that 'the only common property which all commodities have is that they are products of labour' (Bottomore, 1991, p. 565). In addition, Marx argues, in the capitalist mode of production 'all the different types of labour producing commodities' become homogeneous labour (which he refers to as 'abstract labour') in the process of exchange (*ibid.*). While the exchange value of commodities is the form in which value appears, *value* itself is the abstract labour that has been objectified in the commodity; a commodity thus embodies 'a use value and value' rather than 'a use value and an exchange value' (Bottomore, 1991, p. 565).[2] It is important not to attribute 'exchange value' to the commodities themselves because value is created in the production process, not in the exchange process (Fine and Saad-Filho, 2016). As Fine and Saad-Filho (2016, p. 33) explain: 'for surplus value production at least one commodity must contribute more labour time (value) to the outputs than it costs to produce as an input: therefore, one of its use values is the production of (surplus) value', and this commodity is labour power.

The production of surplus value involves an exploitative *social relation* that is not immediately obvious and 'occurs behind the backs of the participants, hidden by the façade of free and equal exchange' (Bottomore, 1991, p. 183). While the wage paid to workers is equal to the value of the commodity they are selling (i.e., their labour power), their control of the production process enables capitalists to extract more labour time in exchange for that wage than it would take the workers to secure their own subsistence (Fine and Saad-Filho, 2016).[3] To summarise, the labour theory of value contends that labour is the source of both value and surplus value (profit) in the capitalist mode of production.

Having examined the source of value and surplus value in capitalist societies, we can return to Marx's explanation of capital as 'self-expanding value'.

Capital as self-expanding value

Marx distinguishes between simple commodity exchange and capitalist exchange (Fine and Saad-Filho, 2016). Money, which acts as a measure and store of value in capitalist societies, is also a means of circulation that facilitates the exchange of commodities. Simple commodity exchange involves exchanging a commodity for money and then exchanging that money for another commodity desired by the person making the transaction. Simple commodity exchange can be depicted as beginning with a commodity (C), selling it for money (M) and using the money to buy a commodity of a different sort (C): C-M-C (with both commodities having the same value). In capitalist exchange, however, the circulation process begins with money acting as capital that buys the inputs needed (such as the means of production and labour power) to produce commodities that will be sold at a profit, and ends with M' (with the difference between M and M' being the profit, or surplus value): M-C-M' (Fine and Saad-Filho, 2016, pp. 30–31). The reproduction of capitalist society involves the constant repetition of this cycle of exchange, production and realisation of profit in the process of capital accumulation. The process of accumulation begins with what Marx refers to as 'primitive accumulation', whereby various mechanisms are used to facilitate a 'transformation of the relations of production' in order to create a working class.[4] Once initiated, the process of accumulation becomes the driving force of the capitalist mode of production; the drive to accumulate capital exists independently of individual capitalists' preferences or desires if they are to retain their existence as capitalists, and this is why Marx describes capital as *self-expanding value* (Bottomore, 1991; Malm, 2013).

Lenin's theory of imperialist rivalries and war

As self-expanding value, capital recognises no physical boundaries, including the boundaries presented by artificially created national borders. Building on Marx's theory of accumulation, Lenin's theory of imperialism analyses self-expanding value 'in the context of a world market created by that accumulation' (Bottomore, 1991, p. 252). Lenin refers to this phase of capital's expansion as the era of 'monopoly capitalism' because it is characterised by an increasing centralisation of production and distribution and a merger of banking and industrial capital (*ibid.*) in the form of finance capital (Bottomore, 1991, p. 200). Competition between nationally based capitals for resources and markets 'divides the world into spheres of influence' and, once the process of this division is completed, establishes the foundations for 'a future inter-capitalist struggle to re-divide the world' (Bottomore, 1991, p. 252).

Adrian Budd (2013) argues that while the imperialist rivalries that led to the catastrophic death and destruction of the two world wars in the first half of the twentieth century did not end, their form and structure changed after World War II. As discussed in the previous chapter, in the aftermath of this war the US and its allies established a web of mechanisms and institutions to expand the geographical scope of capitalist markets and to create new opportunities for capital accumulation (by, for example, instituting policies that promote the privatisation of economic sectors previously owned by national governments). The expansion of capitalist relations of production entails the use of increasing amounts of energy, which increases GHG emissions (UNEP, 2019), despite the urgent need to rapidly and drastically reduce these emissions if we are to limit global warming to below 2 °C (IPCC, 2018). The use of increasing amounts of GHG-emitting fossil fuels to generate the required energy continues *despite* the availability and cheapness of renewable sources of energy – a situation that is perplexing and warrants further investigation.

Andreas Malm's theory of 'fossil capital'

Andreas Malm (2013, p. 17) set himself the task of investigating the origins of what he refers to as the 'fossil fuel economy', which he describes as being 'characterised by self-sustaining growth predicated on growing consumption of fossil fuels, and therefore generating a sustained growth in emissions of carbon dioxide'. Like others before him, he traces the origins of the widespread use of fossil fuels in industrial energy generation to the adoption of machines powered by James Watt's steam engine in nineteenth-century Britain. Malm's investigations, however, also uncover evidence that solves a long-standing puzzle: why the steam engine, which was patented in 1784, was only widely adopted by the owners of the British cotton factories in the mid-1820s (Malm, 2013). Until the mid-1820s, most of the cotton mills in Britain were powered by water mills, which were both cheaper and 'at least as powerful' as steam-powered engines (*ibid.*, p. 31). Reviewing the evidence available in the historical records, Malm concludes that:

> The steam engine did not open up new stores of badly needed energy so much as it gave access to exploitable labour. Fuelled by coal instead of streams, it untied capital in space, an advantage large enough to outdo the continued abundance, cheapness, and technological superiority of water.
>
> (Malm, 2013, pp. 33–34)

He argues that at this point in the development of capital, 'fossil fuels become a necessary material substratum for the production of surplus-value' (Malm, 2013, p. 51). This is why Malm (2013, p. 52) refers to capital as 'fossil capital', which he defines as '*self-expanding value passing through the metamorphosis of fossil fuels into CO2*' (*ibid.*; emphasis in original).

Proceeding to describe fossil capital as 'a triangular relation between capital, labour and a certain segment of extra-human nature', as well as a 'process … of successive valorisations, at every stage claiming a larger body of fossil energy to burn' (*ibid.*, p. 52), Malm modifies Marx's circuit of capital accumulation to derive a 'general formula of fossil capital' (refer to Malm, 2013, pp. 51–52 for the formula). He suggests that the theory of fossil capital, which emphasises the way in which fossil fuels facilitate capital's ability to mitigate the vagaries of nature (such as when the wind blows and when the sun shines) and to continue to control the labour process (and thus workers), might 'provide insights into the stalled shift to renewable energy sources' despite evidence 'that all energy in the world could be provided from wind, water and solar power at little or no extra total cost' (Malm, 2013, p. 60).

Environmental relations in the capitalist mode of production

Malm's work provides a powerful illustration of how to implement Fine and Saad-Filho's (2016, p. 169) advice that the environment 'should be understood in terms of environmental relations (and corresponding structures and conflicts) characteristic of capitalism' rather than as a 'trans-historical conflict between humans and ecological systems or between the environment and the economy'. As demonstrated above, the current ecological crisis reveals a major internal contradiction that capitalism cannot overcome: the reproduction of the capitalist mode of production relies on infinite 'economic growth' while we live on a planet with finite resources and a fragile biosphere that deteriorates as capital expands.

This internal contradiction reveals that the most significant outcome of the global expansion of capital that commenced in the immediate aftermath of World War II and gained momentum in the 1970s/1980s is that it has brought us to the precipice of several interconnected and cascading ecological and social disasters, as encapsulated in the concept of 'the Anthropocene'. While several theorists in the late 1970s and early 1980s argued that existing theoretical perspectives in IR were inadequate analytical tools for understanding and explaining the post-World War II changes in the global political economy (as traditionally defined in the Holocene), the concept of the Anthropocene had not yet been invented. This is perhaps why many GPE analysts generally continued to relegate 'the environment' to a background issue, focusing instead on the issues that were central to the emerging debates about the nature of the phenomena variously referred to as 'globalisation', 'neoliberalism' or 'neoliberal globalisation'.

'Neoliberal globalisation': an essentially contested concept

As noted in the previous chapter, the post-1970s modifications to the global political economy aimed to expand both the operations and the intensity of US-led global capitalism. While concepts such as 'globalisation' and 'neoliberalism' have been used to encapsulate some of these modifications, there are many contestations about the validity, meaning and utility of these terms. One important debate about 'globalisation' is, as McMichael (2000) suggests, the extent to which it is a conscious policy implementation (a 'project') as opposed to a 'natural' occurrence of the operations of a 'free market' (a 'trend'). If one examines the phenomenon closely, it is clear that it is both. 'Neoliberal globalisation' is a project insofar as key players such as economists, think tanks, and policymakers make decisions, formulate policies and institute oversight and compliance instruments to achieve their goals, which amount to facilitating continued (and accelerating) capital accumulation. If, on the other hand, one thinks about capital as self-expanding value, the modifications to the global political economy represent a 'trend' which is inherent in the system. It is therefore not an 'either/or' situation; the best way to understand 'neoliberal globalisation' is as a continuation of the *social* relations that define the capitalist mode of production, which are to be understood as unfolding dynamically. In contrast to this view of the nature of 'neoliberal globalisation', many theorists seem to approach it as if it were a problem that can be 'solved' by calls to somehow reform the global political economy and return to a prior stage of capitalist development (for example, the 'embedded liberalism' stage discussed in the previous chapter, or even Roosevelt's 1930s 'New Deal' stage of US capitalism).

Other analysts adopt a more realistic approach, emphasising the 'project' interpretation without calling for a return to any previous 'golden age' of capitalism, and this approach has the potential to open up avenues for real change rather than joining the futile chorus demanding the 'reform' of capitalism. For example, Douglas cautions against seeing 'neoliberal globalisation' as a *fait accompli*, pointing out that 'by internalising the discourse of "the global", and its associated myths, we all become "vectors" ensuring the transmission of the new normalcy' (Douglas, 1997, p. 174). One way of avoiding the pitfall of becoming a 'transmission vector', and thereby confirming the inevitability of the status quo, is to think of 'neoliberal globalisation' as a process, a part of an incomplete project that manifests as a variety of regulatory experiments which are path-dependent and necessarily evolve 'unevenly across places, territories and scales' (Brenner, Peck and Theodore, 2010, p. 331) and are therefore open to contestation. These authors usefully describe the neoliberalisation process as:

> one among several tendencies of regulatory change that have been unleashed across the global capitalist system since the 1970s: it prioritizes market-based, market-oriented, or market-disciplinary responses to regulatory problems; it strives to intensify commodification in all realms of social life; and it often mobilizes speculative financial instruments to open up new areas for capitalist profit-making.
>
> (Brenner, Peck and Theodore, 2010, p. 329)

A significant outcome of the emergence of this neoliberalising process was the emergence of GPE as a distinct field of academic study.

The Method of Historical Structures: a paradigm shift in IR

The development of GPE in the 1980s can be described as a paradigm shift in IR as it originated when some theorists found the analytical tools provided by IR's orthodox theoretical perspectives, realism and liberalism, inadequate to the task of analysing and explaining the shifts evident in the post-World War II global political economy after the 1970s (Cox with Schechter, 2002). It is important to note, however, that the claim that GPE's multidisciplinarity and global focus was something 'new' is rightfully challenged. Selwyn (2015) points out that eighteenth- and nineteenth-century political economy was always both international and multidisciplinary in its scope. Moreover, he adds, the fragmentation of political economy into the separate discipline areas of economics, political science (and later into IR and IPE) dates back to the 1870s and was at least partially politically motivated: 'The process of intellectual and academic fragmentation was driven, in part at least, by a reaction to the intellectual threat of Marxism and the political threat of emerging European working classes' (*ibid.*, p. 517). GPE theoretical responses to the modifications of the 1970s that paid more attention 'to the economic foundations of power' and initiated 'a critical investigation into change in historical structures' (Cox with Schechter, 2002, p. 79) were not, therefore, entirely novel.

Notwithstanding the long history of Marxist concerns with precisely these issues, Cox's (1981) argument that neither the traditional IR perspectives nor their international political economy (IPE) modifications provide the tools required to critically analyse the nature of, tensions between and contradictions engendered by the social forces underlying the shifts in the global political economy is valid. Citing these reasons, Cox (1981) developed the Method of Historical Structures (MHS) as an analytical tool better suited to that task. Importantly, Cox's critical research agenda aimed to identify the contradictions revealed by the shifts in the prevailing world order with the purpose of exploring whether (and how) the tensions that these contradictions generate between different social forces could lead to emancipatory projects and, ultimately, to a more socially just world order. The focus on emancipatory projects and social justice aligns with one of Cox's key contributions to GPE studies: the distinction he makes between 'problem-solving theory' and 'critical theory'.

Cox's distinction between problem-solving theory and critical theory

Cox sees the purpose of problem-solving theory as being to make existing social and power 'relationships and institutions work smoothly by dealing effectively with particular sources of trouble' (Cox, 1996b, p. 88). By this definition, problem-solving theory can be understood as theory that, while reformist, ultimately supports the status quo. An example of problem-solving theory in the context of the climate crisis would be analyses that promote technology-focused solutions – such as a shift to using renewable sources of energy – while ignoring the wider context of the capitalist relations of production (and the inequalities inherent in these relationships) that all technologies are currently developed and deployed within. As discussed in more detail in Chapter 5, ecosocialists and many climate justice advocates refer to these sorts of solutions as 'false solutions' because, even if they are successful in mitigating anthropogenic climate change, they do not do this in equitable and socially responsible ways that promote climate justice.

In contrast to the reformist aims of problem-solving theory, Cox identifies the purposes of critical theory as being to determine the origins of existing institutions and social power relationships and to identify their internal contradictions and possible future trajectories with the aim of aiding the social forces whose emancipatory struggles might lead to alternative, more just world orders (Cox, 1996b). Ecosocialist theory (which is discussed in Chapter 3) is an excellent example of critical theory applied to the issue of the climate crisis, as it explains why capitalism cannot be reformed to mitigate it (particularly not in ways that are socially just) and argues that the only real solution is 'system change'.

In concluding the discussion on this distinction, it should be noted that Cox does not see 'problem-solving' and 'critical' theory as necessarily mutually exclusive; the distinction he makes between them is intended to highlight their theorists' different normative choices, with the former supporting the status quo and the latter challenging it. As Cox (1996b, p. 91) points out, all sophisticated theories 'usually share some of the features of both problem-solving and critical theory but tend to emphasize one approach over the other'. As signalled at the beginning of this chapter, ontological and epistemological assumptions are important in this respect, and Cox's practical aims of developing theory that facilitates transformative change could not be achieved without challenging the epistemological and ontological assumptions of orthodox IR and IPE perspectives.

The neo-Gramscian GPE perspective: a Method of Historical Structures

The lasting legacy of Robert Cox's theoretical contributions in the field of critical GPE is celebrated in a special issue of the academic journal *Globalizations* (Volume 13, Number 5) published in 2016. It is in this issue that one of Cox's colleagues, Timothy Sinclair, provides a concise summary and critique of, as well as a modification to, Cox's Method of Historical Structures (MHS). Cox introduced the MHS in a 1981 paper entitled 'Social Forces, States, and World Orders: Beyond International Relations Theory', the first of two journal articles that many theorists recognise as having been groundbreaking in the field of critical GPE. Important achievements of Cox's MHS include the critiques of, and the alternatives it offers to, traditional IR ontological assumptions. The ontological assumptions of traditional IR perspectives that Cox critiques include the separation of 'politics' ('the state') from 'civil society' (the economy and other aspects of social life), which enables them to treat a reified, ahistorical 'state' as the basic unit of analysis in the 'inter-state system' without recognising and accounting for the complexity of the real world (Sinclair, 2016). Traditional IR theorists do not, for instance, consider the variety of actors and plurality of forms of state that exist, or the power relations, interrelationships and shifting dynamics between all these actors.

Sinclair's summary of the five main features of Cox's alternative and groundbreaking approach is worth quoting at length because of the brevity and clarity with which it outlines Cox's MHS.

> First, action takes place within a 'framework for action' which limits and constitutes the world. Understanding this requires historical study. Second, theory is also shaped by this framework, in the sense that theorists must be aware of theory's historical character and the continual need for its adjustment as the world changes. Third, the 'framework for action' necessarily changes and the main task of critical theory is understanding

this change. Fourth, the framework for action is an historical structure or combination that brings together thought, material conditions, and institutions. An historical structure does not determine action but 'constitutes the context' within which action takes place. Cox's structure can be read as a constraint but also more actively (but less clearly) as constituting action; so the historical structure does more than limit pre-given agents. Last, frameworks for action, or historical structures, should not be considered in terms of their need for equilibrium maintenance, but more dynamically, in terms of identifying the contradictions and conflicts within them which create the possibility for transformation of the framework for action.

(Sinclair, 2016, pp. 512–513)

Sinclair proceeds to describe the two 'elements' that make up the MHS, the first of which is the existing 'historical structure' that provides the backdrop to the terrain of social action. The components of historical structures are depicted graphically as a 'triangle' of what Cox refers to as 'Forces' and comprise 'material capabilities, ideas, and institutions', as shown in Figure 2.1.

When using this framework of analysis, it is important to constantly keep in mind the implication of the arrows in Figure 2.1; material capabilities, institutions and ideas are all interrelated and interact with, and influence, one another. It is also important to be constantly aware that there is no 'ideal' or eternal 'historical structure'; all historical structures are both distinct and dynamic, and their features can only be revealed through empirical study. The dynamic features of historical structures also mean that history is never 'settled'; historical structures contain contradictions that give rise to 'rival structures' (Sinclair, 2016, pp. 512–513). Furthermore, in Cox's schema the contending forces within any given historical structure can be analysed by applying the balance of the forces evident at any point in time in the historical structure to the second element of the MHS, the three 'levels' or 'spheres of activity'.

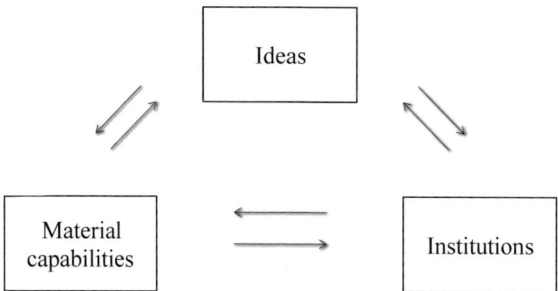

Figure 2.1 Cox's MHS Forces.

Source: Sinclair, T.J. (2016). Robert W. Cox's Method of Historical Structures Redux. *Globalizations*, 13(5), 510–519, publisher Taylor & Francis Ltd, www.tandfonline.com. Reprinted by permission of the publisher.

The 'spheres of activity' are graphically depicted as another triangle, as shown in Figure 2.2, and comprise 'social forces', 'forms of state' and 'world orders'. Cox (1981, 1996b, p. 100) defines the spheres of activity as fluid, and as emerging from the following processes:

> (1) organization of production, more particularly with regard to the *social forces* engendered by the production process; (2) *forms of state* as derived from a study of state/society complexes; and (3) *world orders*, that is, the particular configurations of forces which successively define the problematic of war and peace for the ensemble of states

Again, Cox (1981, 1996b, pp. 100–101) emphasises the way in which the three levels at which social action occurs interact with and influence one another, as well as drawing attention to the significance of the arrows in Figure 2.2.

> The three levels are interrelated. Changes in the organization of production generate new social forces which, in turn, bring about changes in the structure of states; and the generalization of changes in the structure of states alters the problematic of world order … The relationship among the three levels is not … simply unilinear. Transnational social forces have influenced states … Particular structures of world order exert influence over the forms which states take … Forms of state also affect the development of social forces through the kinds of domination they exert.
>
> (Cox, 1996b, pp. 100–101)

What emerges is an analytical schema that seems more able to account for both the complexity and dynamic nature of the global political economy and to incorporate a much wider range of actors than traditional IR and dominant IPE perspectives provide tools for. Importantly, it is also a schema that opens up possibilities for change, as is evident when Cox (1996b, pp. 116–117) expands on his decision to refer to 'world orders' as opposed to

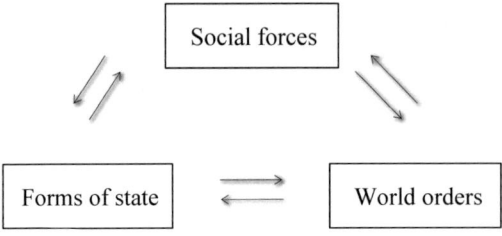

Figure 2.2 Cox's MHS Spheres.

Source: Sinclair, T.J. (2016). Robert W. Cox's Method of Historical Structures Redux. *Globalizations*, 13(5), 510–519, publisher Taylor & Francis Ltd, www.tandfonline.com. Reprinted by permission of the publisher.

alternative terms such as 'inter-state system' or 'world system' with reference to how the term 'inter-state' refers only to periods in history when 'states' exist, whereas 'world orders' always exist. His preference of the word 'order' instead of 'system' is for similar reasons; the word 'system' has connotations of equilibrium, whereas the sense in which he uses 'order' is to describe 'the way things usually happen (*not* the absence of turbulence)' (*ibid.*).

Sinclair's 'Method of Historical Structures Redux'

While acknowledging the robustness of Cox's MHS as an analytical schema that is well equipped to deal with the dynamic complexities of the real world as well as the challenges that these complexities present to actors trying to create socially just alternative futures, Sinclair (2016) also addresses some critiques levelled against it and proposes modifications to address these critiques. After summarising criticisms related to some sources of confusion regarding the application of the two 'triangles', ontological choices and the role of agency in the original analytical framework developed by Cox, Sinclair suggests a few modifications that address these issues and are designed to strengthen the analytical potential of the MHS. The 'MHS Redux', as Sinclair refers to the modified analytical framework, is graphically depicted as a triangle of 'Forces Redux' (refer to Figure 2.3) and a diamond of 'Spheres Redux' (refer to Figure 2.4), with the different shapes helping to eliminate one of the sources of confusion confronted by researchers who apply the methodology for empirical research: having to make sense of how the two triangles in Cox's MHS relate to each other.

As is evident in Figure 2.3, Sinclair makes two modifications to Cox's original schema of 'forces'. He explicitly adds 'reproductive capabilities' to Cox's original 'productive capabilities' to address concerns about ontological choices that neglect gender and reproduction, and he disaggregates Cox's 'ideas' into 'competing ideas' and 'social facts', amalgamating the latter with

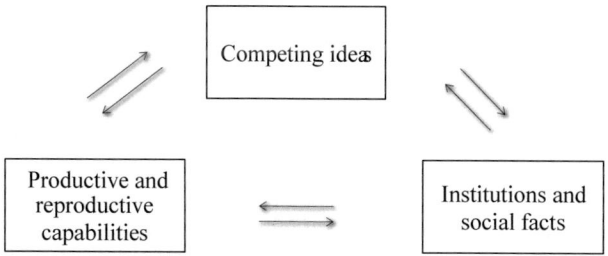

Figure 2.3 Sinclair's MHS Forces Redux.

Source: Sinclair, T.J. (2016). Robert W. Cox's Method of Historical Structures Redux. *Globalizations*, 13(5), 510–519, publisher Taylor & Francis Ltd, www.tandfonline.com. Reprinted by permission of the publisher.

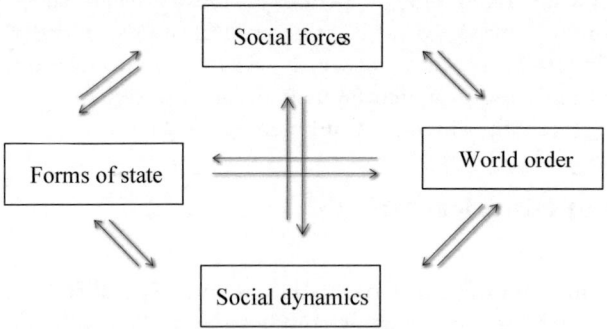

Figure 2.4 Sinclair's MHS Spheres Redux.

Source: Sinclair, T.J. (2016). Robert W. Cox's Method of Historical Structures Redux. *Globalizations*, 13(5), 510–519, publisher Taylor & Francis Ltd, www.tandfonline.com. Reprinted by permission of the publisher. This figure is an amended version of the original figure and was developed in collaboration with the author, T.J. Sinclair.

Cox's 'institutions', Sinclair's rationale for the second modification is that since Cox defines 'institutions more broadly than mere organizations', the 'norms and assumptions [that] structure our lives well away from … competing ideas,' that is, 'social facts', can be added to this category (Sinclair, 2016, pp. 514–517). This allows Sinclair to isolate 'competing ideas', which are collectively held ideas that conflict with 'social facts', in their own category. The Marxist notion of 'working-class consciousness' is a good example of a 'competing idea'.

Sinclair's modification of Cox's second triangle of 'spheres' (Figure 2.4) is more extreme, transforming Cox's triangular shape into a diamond by adding a fourth sphere called 'social dynamics' to complement Cox's 'narrow understanding' of 'social forces' as those forces 'engendered by the production process'. 'Social dynamics' are designed to encapsulate 'the vast range of human conflict and cooperation not reducible to production. Struggles by social movements about things such as human rights and the biosphere are manifestly consequential and need to be recognised as such' (Sinclair, 2016, p. 515).

Having modified Cox's original framework, Sinclair (2016) clarifies how the two schemas 'fit together' and can be used by researchers analysing aspects of social reality.

> The first, the forces, is the static or synchronic understanding of how things fit together (e.g. Thatcherism) while the spheres … allow for understanding of the broader context and incorporates [sic] potential contradictions between elements (e.g. Thatcherism vs. Soviet Communism, for example).
>
> (Sinclair, 2016, p. 518)

This framework could also facilitate an analysis of how the neoliberalising forces dominating the current historical structure generate contradictions that manifest as the organic crisis of global capitalism and provoke opposition in various quarters, including from the 'alter-globalisation' global justice movement that gained prominence with the 1999 mass demonstrations in Seattle against the World Trade Organization (WTO) and other economic institutions actively promoting the neoliberal agenda (Graeber, 2013).

The MHS and MHS Redux as heuristic devices: strengths and limitations

The strengths of the MHS and the MHS Redux become evident when one considers the main purpose of critical theory in the social sciences: to aid the sort of empirical analysis that can be used as 'a guide to strategic action for bringing about an alternative order' (Cox, 1996b, p. 90). This task requires an analytical framework that can accommodate change by allowing for its heuristic application rather than a theoretical perspective that assumes universal 'laws' that, whether wittingly or not, support the status quo by implying a static, immutable 'present' that cannot be challenged. Humphreys (2010) identifies three main functions of a theoretical perspective deployed as an heuristic device.

> First, it indicates what sort of explanation is required, for example by suggesting a particular level of analysis or a causal rather than an interpretive approach. Second, it provides the conceptual categories that are used to navigate through and to organize empirical material. Third, it indicates what mechanisms, actors, chance factors and background conditions are worth examining.
>
> (Humphreys, 2010, p. 263)

According to these criteria, the MHS and MHS Redux are powerful heuristic devices that facilitate analyses at a variety of levels (local, state and global) using a range of conceptual categories (material capabilities, institutions and ideas – all of which can be examined using the Marxist concepts defined in the first part of this chapter and the Gramscian concepts defined later in this chapter). The arrows in the MHS and MHS Redux schema also remind the analyst to consider the roles of, and the relationships between, a variety of actors (ranging from those who currently have much economic and political power to those who are engaged in struggles against this power: the working class, subaltern groups and social movement actors).

Another measure of the utility and strength of heuristic devices that Humphreys (2010) identifies is the extent to which they can be used to guide empirical research on the most important issues of the time. The central problem within IR when Cox developed the MHS related to the issue of geopolitical instability as a result of the potential weakening of US hegemony and, most importantly, the possibilities this instability raised for its replacement by more progressive alternative world orders. As Cox explains:

> Since the practical issue at the present is whether or not the *pax Americana* has irretrievably come apart and if so what may replace it, two specific questions deserving attention are: (1) what are the mechanisms for maintaining hegemony in this particular historical structure? And (2) what social forces and/or forms of state have been generated within it which could oppose and ultimately bring about a transformation of the structure?
>
> (Cox, 1996b, p. 107)

While these were, indeed, the most important known challenges facing both progressive theorists and oppositional groups at the time that Cox was writing, it is here argued that in addition to these issues, which remain central to how the Anthropocene unfolds (Klein, 2014), the general deterioration of the biosphere and the effects of climate change require that these ecological issues should be given the prominence that the severity of their consequences deserves in all the work we now do as social scientists. As noted in Chapter 1 when discussing the Anthropocene, Clive Hamilton maintains that addressing these challenges requires a shift to 'Anthropocene thinking' in academics' research focus. This call for a shift in thinking also aligns with Cox's belief that 'theorists must be aware of theory's historical character and the continual need for its adjustment as the world changes', thereby necessarily changing the 'framework for action' which it is 'the main task of critical theory' to understand (Sinclair, 2016, p. 512).

While current understandings encapsulated in the concept of the Anthropocene have changed the 'framework for action' in the real world, Cox's MHS, which treats 'nature' as a background category within the concept of 'material capabilities', has yet to catch up with this development. This is not surprising, as the MHS was developed a few years before US scientist James Hansen's 1988 congressional testimony on AGW that generated a wider public awareness of the severity of its consequences. In addition, the Anthropocene concept was also not available to Robert Cox in the 1980s. It was only subsequent scientific research in Earth System science (a field of study developing concurrently with critical GPE) that raised concerns about the dangers of overshooting planetary boundaries to the extent that we cross 'tipping points' in the Earth System. The warnings that crossing such tipping points could propel the planet into new and unpredictable states that are less than ideal for the life-forms (including humans) that currently inhabit it (Steffen et al., 2015) were not available to Cox. These dangers imply, however, that the most important challenges we face today revolve around the maintenance of a habitable planet. There is evidence that Cox would concur with this conclusion; in his later writings, he describes the biosphere as constituting 'the material, ecological basis of human life', and he also emphasises the urgency of addressing threats to its integrity (Cox, 2007, p. 526; see also Cox with Schechter, 2002).

Sinclair (2016) also explicitly refers to the importance of environmental issues, justifying his decision not to create a separate category for 'the environment' by agreeing with Cox that the changes required to protect the integrity of the biosphere will necessarily be achieved through social struggles. As Sinclair (2016, p. 515) puts it, 'while we are material beings as well as social ones, the struggle over how to address the environment is part of the social dynamic, and how we resolve it will not be reducible to material necessity alone'. Given the evident inability of capital to solve the problem of the degradation of the biosphere, which Marxists and ecosocialists argue results from fundamental contradictions inherent in the capitalist mode of production, Cox and Sinclair are correct to argue that the issue of whether or not the planet remains habitable will only be resolved through social action. However, it can nevertheless be argued that both the MHS and MHS Redux lack an appropriate analytical category to account for the biosphere's 'agency', and the MHS Redux II attempts to address this gap.

MHS Redux II: an analytical tool for the Anthropocene

Following the understanding that 'Cox's categories are not exhaustive as he sets them out in 1981, but are open to redefinition and adaptation over time' (Sinclair, 2016, p. 514), the MHS Forces Redux II introduces a new category, 'the biosphere', to Sinclair's MHS Forces Redux. The IPCC defines the biosphere as:

> The part of the earth system comprising all ecosystems and living organisms, in the atmosphere, on land (terrestrial biosphere) or in the oceans (marine biosphere), including derived dead organic matter, such as litter, soil organic matter and oceanic detritus.
>
> (IPCC, 2014, p. 1254)

While it would be most accurate to enclose the entire MHS Redux within an all-encompassing umbrella category called 'Earth System', this would not aid systematic analyses. Such a modification would, in fact, risk being interpreted as following in the footsteps of the 'new materialists', who draw 'upon the insights of science to offer a new, "flat" ontology that emphasises the agency of all things' and thereby obscure rather than reveal existing power structures (Choat, 2018, p. 1031; refer also to Malm, 2019 and to Hamilton, 2017, chapter 3). Choat points out that such flat ontologies, which 'entangle' human and natural agents so that they become indistinguishable, risk 'depoliticising situations':

> if we do not know which actors are more important than others, then we deny ourselves the ability to intervene in the hope of altering the existing balance of forces. Indeed, more than simply leaving us without the resources to analyse and resist asymmetrical power relations, in its

emphasis on relations of becoming, contingency, change, openness and suppleness, new materialism tends to obscure the very existence of enduring or rigid structures of power and the reproduction of relations of domination and exploitation.

(Choat, 2018, p. 1036)

To avoid this situation, while still developing an analytical category that gives the biosphere 'agency' and places it at the centre of research concerns, the MHS Redux II modification encloses Sinclair's 'productive and reproductive capabilities' within an 'umbrella' category, the 'biosphere' (refer to Figure 2.5).

In the MHS Redux II, the biosphere's 'agency' is strictly limited and does not allow for intentionality on the part of the 'agent'; the biosphere's 'agency' comprises of natural processes that react, strictly according to the laws of physics, to natural forces that are either intentionally or unintentionally unleashed by the actions of human agents. This approach is designed to facilitate analyses that, while incorporating the causes and effects of 'natural phenomena' (which are now modified by the effects of human activities), can still be conducted using Marx's notion of historical materialism. Unlike the 'new materialisms', Marxist historical materialism:

seeks not to (re)define matter but to interrogate the historically specific material conditions of human production and reproduction, and hence the *material conditions* of the development and uses of science, the production and role of objects and agents, and our labour within and upon nature.

(Choat, 2018, p. 1028, emphasis in original)

As well as constituting a more detailed depiction of the real world because the state of the biosphere necessarily affects the conditions under which both productive and reproductive activities occur, the category 'biosphere' introduces an important dynamic element within what could previously be described as 'static forces', at least for purposes of empirical analysis at any specific point in time.

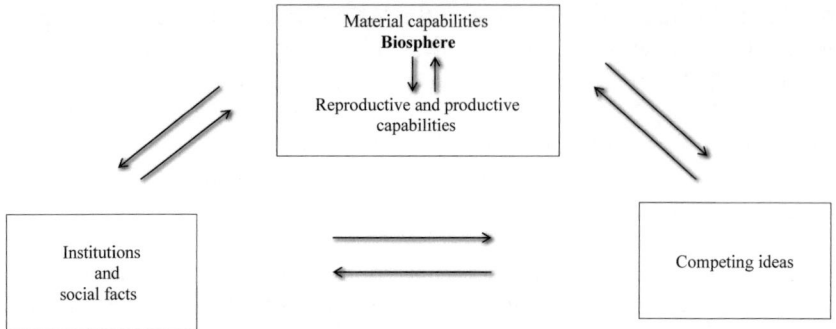

Figure 2.5 MHS Forces Redux II.

Changes in the biosphere are not only currently imperfectly understood but, given its complexity and the way in which its different elements interact and influence one another, its future trajectory is also, in principle, impossible to predict. Hamilton, who has for many years been warning about the hubris of scientists who believe that they can 'geo-engineer' our planet and control its climate (Hamilton, 2013), cautions that 'in the transition from the Holocene to the Anthropocene new forces have been unleashed that we can only ever understand imperfectly, and regulate even less' (Hamilton, 2017, p. 70). While these features of the biosphere (both in reality and as an analytical category) add to the already formidable challenges of understanding the nature of the global political economy and identifying rival structures within it, there is much evidence that they are too important to ignore. For example, anthropogenic global warming already impacts human societies in the form of 'extreme weather' events, as discussed in Chapters 1 and 7. Moreover, the effects of 'extreme weather' events such as floods are exacerbated with rising sea levels as land-ice melts at the poles and as ocean waters warm and expand (Hansen et al., 2016). Also important and unpredictable are the social forces that the effects of these 'natural disasters' may unleash. These social forces (particularly the ecosocialist contingent of the climate justice movement) are the subject of the extended case study that constitutes the rest of this book, throughout which the MHS Forces Redux II and the MHS Spheres Redux are applied to critically analyse the tensions, contradictions, and possibilities for transformational change arising from all these developments. This analysis is conducted with reference to a number of Marxist analytical concepts, some of which were introduced earlier in this chapter and others of which were initially developed by Italian Marxist Antonio Gramsci and are highlighted by Robert Cox in his second groundbreaking paper, 'Gramsci, Hegemony, and International Relations: An Essay in Method' (Cox, 1983).[5]

Gramscian and neo-Gramscian analytical concepts

As many Gramscian scholars (for example, Peter Thomas (2010, 2013) and Adam Morton (2007) demonstrate, the meanings of the concepts outlined below and used in this study are both complex and contested. Given the already extensive scope and purpose of this book, however, the approach adopted in the use of selected Gramscian concepts, while drawing on the expertise of Gramscian scholars, is similar to Cox's, who wrote:

> This essay sets forth my understanding of what Gramsci meant by hegemony and these related concepts, and suggests how I think they may be adapted, retaining his essential meaning, to the understanding of problems of world order. It does not purport to be a critical study of Gramsci's political theory, but merely a derivation from it of some [useful] ideas.
>
> (Cox, 1996c, p. 124)[6]

Hegemony is a key concept in traditional realist and liberal IR theories and in Gramsci's work and neo-Gramscian GPE analyses, but IR and GPE theorists define the concept in very different ways. While IR and problem-solving IPE theorists generally refer to 'strong' states as being hegemonic in the 'inter-state system' because their superior military or economic material power allows them to impose policies that suit their own interests on other states, critical GPE theorists such as Cox emphasise 'the consensual aspect' of hege-monic power (Cox, 1996c, p. 127). In the neo-Gramscian GPE literature, therefore, 'hegemony' generally refers to a form of class rule based on consent (Cox, 1981, 1983) that is 'backed up only in the last instance by the coercive apparatus of the state' (Overbeek, 2000, p. 172). Budd (2013), however, presents much compelling evidence to support his argument that not only do neo-Gramscians misinterpret what Gramsci meant when they overemphasise the consensual aspect of hegemony, but they also frequently mistake acquiescence for consent in their historical analyses. The consent that neo-Gramscians refer to derives from the way in which, in specific capitalist societies, there have been historical periods when the ruling class truly represented the 'general interest' of leading subordinate classes or, in Gram-scian terms, *subaltern* groups and classes.[7] In those hegemonic historical periods, which Budd (2013) points out were very short periods of time, the ruling class made important concessions that gave the subordinate groups a stake in the status quo.

Both Gramsci and neo-Gramscian GPE analysts identify the way in which the ruling class is able to exert its influence over prevalent ideologies and ideas of morality through its control over 'the myriad of institutions and rela-tionships in civil society' (Overbeek, 2000, p. 173) as another tool that is used to construct hegemony. Gramsci's notion of the *integral state* or the *extended state* encapsulates this dynamic between the formal apparatus of the state ('politics') and institutions such as schools, religious organisations and the mass media within 'civil society'. The intellectual work of the ruling class' *organic intellectuals*, whose function it is to develop and sustain 'the mental images, technologies, and organizations which bind together the members of a class and of an historic [sic] bloc into a common identity', is crucial in this respect (Cox, 1996a, p. 132) as it helps to construct and maintain the hege-mony of social formations that Gramsci refers to as *historical blocs*.

At the national level, *historical blocs* are created when a fundamental social class succeeds in integrating 'a variety of different class interests and forms of identity within a "national-popular" alliance' (Morton, 2007, p. 97). The extent to which a historical bloc is hegemonic is one important factor shaping the particular form of a state (with examples of different forms of state being 'liberal-democratic welfare states', 'neoliberal competition states' and dictator-ships). Gramsci notes that *international* hegemony also constitutes 'a form of class rule based on consent more than on coercion, and on accommodation of sub-ordinate interests rather than on their repression' (Overbeek, 2000, p. 174). Neo-Gramscian GPE perspectives see hegemony in the global system as a form of class rule that integrates social, economic and political structures (Overbeek, 2000, p. 176). Thus, Cox argues:

> The hegemonic concept of world order is founded not only upon the regulation of inter-state conflict, but also upon a globally-conceived civil society, i.e., in a mode of production of global extent which brings about links among social classes of the countries encompassed by it.
>
> (Cox, 1983, p. 171)

The hegemony of historical blocs is, however, vulnerable and must be continually constructed, maintained and defended 'in the face of constant resistance and pressures' which present opportunities for the formation of *counter-hegemonic movements* that have the potential to challenge the hegemony of the ruling class (Morton, 2007, p. 97). As Gill (1993, p. 44) notes, neo-Gramscian GPE perspectives explain the causes of social crises and transformations with reference to 'the disintegration of social hegemonies, and the formation of counter-hegemony in the global political economy'. Hegemonic historical blocs face their greatest challenges during periods characterised by what Gramsci refers to as an *organic crisis*, which heralds 'relatively long-term and permanent changes, as opposed to "conjunctural" [changes]' (Cox, 1996b, Note 25, p. 120) that are temporary and can be overcome through 'problem-solving'. As Gramscian scholar David Forgacs explains:

> An 'organic crisis' is a crisis of the whole system, in which contradictions in the economic structure have repercussions through the superstructures. One of its signs is when the traditional forms of political representation (parties or party leaders) are no longer recognized as adequate by the economic class or class fraction which they had previously served to represent. It is therefore a crisis of hegemony, since it occurs when a formerly hegemonic class is challenged from below and is no longer able to hold together a cohesive bloc of social alliances.
>
> (Forgacs, 2000, p. 427)

During an organic crisis, 'the structures and practices that constitute or reproduce a hegemonic order fall into chronic and visible disrepair, creating a new terrain of political and cultural contention, and the possibility (but only the possibility) of social transformation' (Carroll, 2010, pp. 170–171).

The agents of social transformation take the organisational form of what Gramsci refers to as 'the *modern Prince*', which is constituted by 'a coalition of the rebellious subalterns, engaged in acts of self-liberation of hegemonic politics' (Thomas, 2013, p. 33). The modern Prince thus refers neither to a 'concrete individual' nor to a 'single centralized entity', but to 'a dynamic collective process' from which a 'distinctively new type' of political party emerges (Thomas, 2013, p. 32). Gramsci's conception of this new sort of political party envisages it as a 'collective organism' that represents 'an expansive revolutionary process in movement' and, rather than solidifying and deforming its development with traditional constitutional forms, it represents 'a pedagogical laboratory for unlearning the habits of subalternity and discovering new forms of conviviality, mutuality and collective self-determination' (Thomas, 2013, pp. 32–33). While organic crises present opportunities for the formation of counter-hegemonic historical blocs by subaltern classes and subaltern social groups working together, whether or not this potential is realised depends on many factors. One crucial variable that affects whether or not a counter-hegemonic bloc can take advantage of an organic crisis to strengthen its position enough to effect real change is how much progress has been made in what Gramsci refers to as the *war of position*.

Gramsci distinguishes between a *war of manoeuvre* (which entails an attempt to overthrow the state or ruling class, as happened in 1917 in Russia when the Bolsheviks overthrew the tsarist government) and a *war of position* (Rupert, 2005). The war of position 'is a strategy for the long-term construction of self-conscious social groups into a concerted emancipatory bloc within society' (Morton, 2007, p. 105). The importance of an effective war of position cannot be overstated: 'It is only when the war of position has built up a combination of organized social forces strong enough to challenge the dominant power in society that political authority in the state can be effectively challenged and replaced' (Morton, 2007, pp. 105–106). For Cox, too, the war of position is essential and constitutes 'a long-term task for organic intellectuals working in constant interaction with the groups whose dissent from the established order makes them candidates for inclusion' (Cox, 1987, p. 390). But, as history shows, the ruling class is adept at coopting key elements of dissenting groups (such as the leaders of trade unions and non-governmental organisations that begin as oppositional and then enter 'mainstream' institutions that give them a stake in supporting and upholding the status quo), a ruling class tactic Gramsci refers to as *trasformismo*.

Another danger to achieving the aims of counter-hegemonic historical blocs is the onset of what Gramsci refers to as *passive revolution*, whereby the war-of-position strategy is stalled, being 'strong enough to provoke opposition, but not strong enough to overcome' those in power (Morton, 2007, p. 106). Thomas (2013, p. 23) explains that:

> In its broadest sense, the notion of passive revolution for Gramsci signified a distinctive process of (political) modernization that lacked the meaningful participation of popular classes in undertaking and consolidating social transformation. On the other hand, having to resort to retaining power by means of passive revolution demonstrates the underlying weakness of the ruling class as 'revolutions from above' indicate a failure to achieve hegemony through consent, which in turn presents opportunities for counter-hegemonic forces to ally in the form of a 'modern Prince' and develop a new kind of consensual politics in the process of engaging in an 'active revolution'.
>
> (Thomas, 2013, p. 30)

In addition to the dangers of passive revolution and *trasformismo*, subaltern classes and social groups trying to build a counter-hegemonic historical bloc face several other formidable challenges as a result of the economic, social and political changes wrought by the neoliberal globalisation project over the past three decades. As Cox warns, the strategic implications of the need to win the war of position before attempting to take state power are 'fraught with difficulties'.

> To build up the basis of an alternative state and society upon the leader-ship of the working class means creating alternative institutions and alternative intellectual resources within existing societies and building bridges between workers and other subordinate classes. It means actively building a counter-hegemony within the established hegemony while resisting the pressures and temptations to relapse into pursuit of incre-mental gains for subaltern groups within the framework of bourgeois hegemony.
>
> (Cox, 1996a, pp. 128–129)

Regardless of the difficulties of the task, it is crucial that the organic intellec-tuals of subordinate groups and classes understand that they need to guide the evolution of 'a clearly distinctive culture, organization, and technique, and do so in constant interaction with the members of the emergent block' (Cox, 1996a, p. 132). In the process of building a new world within the existing one, working-class consciousness must also evolve.

Gramsci identifies three levels of working-class consciousness, beginning with the *economico-corporate consciousness*, which focuses on the material interests of a particular group, such as workers at a particular workplace (Cox, 1996a). At a more advanced level, class consciousness extends to a whole social class and is understood in terms of *class solidarity*, but remains limited in its focus on economic issues. The most advanced level of consciousness, *hegemonic* class consciousness, harmonises the interests of the leading class 'with those of sub-ordinate classes and incorporates these other interests into an ideology expressed in universal terms' (Cox, 1996a, p. 133). In the nineteenth and twentieth centuries, historical accounts suggest that many workers in the advanced capitalist societies had at least the most basic, economico-corporate class-conscious awareness that Gramsci refers to (for example, refer to Hyman, 1975). This situation was reversed during the Reagan and Thatcher eras, with individualism having widely displaced even this elementary-level class consciousness as a result of the weakening of organised labour in advanced capitalist societies and the successes of the ideological work of capitalism's organic intellectuals in vilifying the labour movement (McIlroy, 2011; McIntyre and Hillard, 2012).[8] There is, however, one aspect of people's consciousness that could work in favour of progressive forces: the contemporary wide-ranging *common sense* awareness and discussion about the injustices of 'globalisation', a system that is now widely acknowledged to favour 'elites' ('the 1%') at the expense of 'ordinary people' ('the 99%').[9] Gramsci identifies one crucial task of the organic intellectuals of subordinate groups as being to draw attention to, emphasise and elaborate on the 'critical elements and "good sense" which are already present within people's "common sense"' (Forgacs, 2000, p. 323; see also Hyman, 1975). Because 'common sense' is neither 'monolithic' nor 'univocal', but is rather 'a syncretic historical residue, fragmentary and contradictory, open to multiple interpretations and potentially supportive of very different kinds of social visions and political projects' (Rupert, 2005, pp. 487–488), much of the war of position is fought on this ideological terrain and it is thus crucial to 'engage with what they actually think' if one is to 'shift people's common sense' in progressive directions (Forgacs, 2000, p. 324).

Another important element of the ideological *war of position* is what Gramsci refers to as the *ethico-political sphere*: 'the ideological, moral and cultural cements which bond a society together' (Forgacs, 2000, p. 190). As discussed in later chapters, ethical and moral issues may become more salient as increasing numbers of people and communities suffer as a result of the ecological, economic, social and political crises that are already destabilising many people's lives and are likely to be exacerbated as the effects of AGW continue to manifest in more intense and frequent 'extreme weather' events such as extensive droughts, rising sea levels and other unpredictable changes to the Earth System that affect human productive and reproductive capabilities. While there is as yet little indication that AGW and the onset of the Anthropocene is the cause of much concern (particularly if one considers the outcomes of the 2019 Australian and UK elections), it is entirely plausible that future catastrophic events related to these developments may provoke such concerns – as well as critiques of the socio-economic system that has caused these shifts in the Earth System and a willingness to experiment with other ways of living. To date, elite responses to hardships experienced as a result of 'extreme weather' events have favoured the wealthy and the privileged, and this tendency to ignore the problems of the great bulk of humanity that is neither wealthy nor privileged in the face of 'natural catastrophes' is likely to continue. An awareness of this issue may make people more receptive to questioning not only the ability of capitalism to stop damaging the biosphere that all life-forms are an integral part of and depend on for their survival, but also the morality of capitalist relations of production, so that social justice issues become increasingly important in the war of position that global justice actors and ecosocialists are engaged in as the organic crisis of global capitalism continues to unfold. The next chapter provides an overview of ecosocialist theoretical contributions to this war of position.

Notes

1 Refer to Bottomore (1991) for overviews of some of the controversies regarding the interpretation of Marxist concepts, noting also that new debates about their interpretation and application are continually being published in books and academic journals.

2 Value is measured by the average necessary time it takes to produce any given commodity 'under the conditions of production normal for a given society and with the average degree of skill and intensity of labour prevalent in that society', which Marx refers to as 'socially necessary labour-time' (Marx, cited in Bottomore, 1991, p. 503).

3 The value of the commodity labour power 'typically corresponds to the labour time socially necessary to produce the wage goods regularly purchased by the working class' (Fine and Saad-Filho, 2016, p. 33).

4 Facing 'fierce resistance' that was 'brutally crushed' (Fine and Saad-Filho, 2016, p. 67), in England this involved a prolonged process of state collaboration and the use of legal mechanisms that denied peasants the access they had previously had to land (the dominant pre-capitalist means of production) so that they became reliant on selling their labour power as a commodity in return for a wage (see also Bottomore, 1991).

5 Gramsci's Marxism is emphasised because this is the reading of Gramsci that is relevant to the analyses conducted in this book. As Thomas (2013, p. 28) points out:

> it has often been claimed that Gramsci was fundamentally a theorist of the cultural superstructures, one who was not only a strong critic of economic determinism but perhaps even ignorant of economic theory. Sometimes, it has even been asserted that Gramsci's concept of hegemony represents the beginning of a 'post-Marxism', which logically should reject the Marxist critique of political economy and its emphasis upon class. Such readings, however, neglect the totality of the *Prison Notebooks*, which contain extensive notes dedicated to discussions of Marx's *Capital* and economic history. They also neglect the context of Gramsci's political activism, which remained fundamentally directed against what he repeatedly characterized as the 'dictatorship' of the bourgeoisie, including and especially in its fascist variant.

6 It should be noted, however, that the way some Gramscian concepts are used in this book deviates from standard neo-Gramscian interpretations of these concepts given the convincing critiques of the latter interpretations presented by various Gramscian scholars (see, for example, Budd, 2013).

7 Thomas (2013, p. 32) points out that the concept of 'subaltern social groups' constitutes a 'novel addition to Marxist class analysis' and 'is not limited to the classes exploited in the capitalist labour process, but includes all social groups oppressed and consigned to the "margins" of history'. Green's (2011) reading of Gramsci's work also leads him to conclude that 'Ultimately, for Gramsci, subalternity is not merely limited to class relations; subalternity is constituted through exclusion, domination, and marginality in their various forms' (Green, 2011, p. 388).

8 Stuart Hall (2011) provides a fascinating account of the 2011 'London Riots' sparked by the death of Mark Duggan, who was shot by police falsely claiming he was threatening them. Duggan's killing provoked public protests that soon escalated into riots across a number of London boroughs and lasted for nearly a week (6–11 August 2011). Subsequent research shows that the participants in these riots were mostly poor, unemployed and young, and Hall explains their motivations with reference to the prevailing ideologies of individualism (that tears the social fabrics of communities), and material consumerism (which urges 'consumers' to measure their own and others' worth in terms of how much 'stuff' they have and how many brand-name products they own). Keaney (2014, p. 62) refers to the London 'rioters' as 'the fruit of the deindustrialization of the 1980s and the closing of opportunities for semi-skilled and unskilled labor outside of low-paying jobs in the retail and ancillary service sectors'.

9 As discussed in more detail in Chapter 7, 'We are the 99%' was the popular slogan of the 'Occupy Movement' that spread around the capitalist world in 2012 in reaction to government responses to the GFC (Graeber, 2013).

References

Bottomore, T. (ed.) (1991). *A Dictionary of Marxist Thought* (2nd edn). Oxford: Blackwell.

Brenner, N., Peck, H. and Theodore, N. (2010). After Neoliberalisation? *Globalizations*, 7(3), 327–345.

Budd, A. (2013). *Class, States and International Relations: A Critical Appraisal of Robert Cox and Neo-Gramscian Theory*. London: Routledge.

Carroll, W.K. (2010). Crisis, Movements, Counter-Hegemony: In Search of the New. *Interface: A Journal for and about Social Movements*, 2(2), 168–198.

Choat, S. (2018). Science, Agency and Ontology: A Historical-Materialist Response to New Materialism. *Political Studies*, 66(4), 1027–1042.

Collier, D., Hidalgo, F.D. and Maciuceanu, A.O. (2006). Essentially Contested Concepts: Debates and Applications. *Journal of Political Ideologies*, 11(3), 211–246.

Cox, R.W. (1981). Social Forces, States, and World Orders: Beyond International Relations Theory. *Millennium: Journal of International Studies*, 10(2), 126–155.

Cox, R.W. (1983). Gramsci, Hegemony and International Relations: An Essay in Method. *Millennium: Journal of International Studies*, 12(2), 162–175.

Cox, R.W. (1987). *Production, Power and World Order: Social Forces in the Making of History*. New York: Columbia University Press.

Cox, R.W. (1996a). Gramsci, Hegemony, and International Relations: An Essay in Method (1983). In Cox, R.W. with Sinclair, T.J., *Approaches to World Order*. Cambridge: Cambridge University Press, 124–143.

Cox, R.W. (1996b). Social Forces, States, and World Orders: Beyond International Relations Theory (1981)'. In Cox, R.W. with Sinclair, T.J., *Approaches to World Order*. Cambridge: Cambridge University Press, 85–123.

Cox, R.W. (1996c). 'Take Six Eggs': Theory, Finance, and the Real Economy in the Work of Susan Strange (1992). In Cox, R.W. with Sinclair, T.J., *Approaches to World Order*. Cambridge: Cambridge University Press, 174–190.

Cox, R.W. (2007). 'The International' in Evolution. *Millennium: Journal of International Studies*, 35(3), 513–527.

Cox R.W. with Schechter, M.G. (2002). *The Political Economy of a Plural World: Critical Reflections on Power, Morals and Civilization*. New York: Routledge.

Cunningham, F. (2002). *Theories of Democracy: A Critical Introduction*. London: Routledge.

de Lucia, V. (2014). The Climate Justice Movement and the Hegemonic Discourse of Technology. In Dietz, M. and Garrelts, H. (eds), *Routledge Handbook of the Climate Change Movement*. New York: Routledge, 66–83.

Douglas, I.R. (1997). Globalisation and the End of the State? *New Political Economy*, 2(1), 165–177.

Fine, B. and Saad-Filho, A. (2016). *Marx's 'Capital'* (6th edn). London: Pluto Press.

Forgacs, D. (ed.) (2000). *The Antonio Gramsci Reader: Selected Writings 1916–1935*. New York: New York University Press.

Gallie, W.B. (1956). Essentially Contested Concepts. *Proceedings of the Aristotelian Society, New Series*, 56 (1955–1956), 167–198.

Gill, S. (ed.) (1993). *Gramsci, Historical Materialism and International Relations*. Cambridge: Cambridge University Press.

Graeber, D. (2013). *The Democracy Project: A History. A Crisis. A Movement*. London: Allen Lane.

Green, M.E. (2011). Rethinking the Subaltern and the Question of Censorship in Gramsci's Prison Notebooks. *Postcolonial Studies*, 14(4), 387–404.

Hall, S. (2011). The Neo-Liberal Revolution. *Cultural Studies*, 25(6), 705–728.

Hamilton, C. (2013). *Earth Masters: Playing God with the Climate*. Crows Nest, NSW: Allen & Unwin.

Hamilton, C. (2017). *Defiant Earth: The Fate of Humans in the Anthropocene*. Crows Nest, NSW: Allen & Unwin.

Hansen, J., Sato, M., Hearty, P., Ruedy, R., Kelley, M., Masson-Delmotte, V., Russell, G., Tselioudis, G., Cao, J., Rignot, E., Velicogna, I., Tormey, B., Donovan, B., Kandiano, E., von Schuckmann, K., Kharecha, P., Legrande, A.N., Bauer, M. and Lo, K-W. (2016). Ice Melt, Sea Level Rise and Superstorms: Evidence from Paleoclimate Data, Climate Modelling, and Modern Observations that 2 °C Global Warming Could be Dangerous. *Atmospheric Chemistry and Physics*, 16, 3761–3812.

Humphreys, A.R.C. (2010). The Heuristic Application of Explanatory Theories in International Relations. *European Journal of International Relations*, 17(2), 257–277.

Hyman, R. (1975). *Industrial Relations: A Marxist Introduction*. London: The Macmillan Press.

Hamilton, C. (2017). *Defiant Earth: The Fate of Humans in the Anthropocene*. Crows Nest: Allen & Unwin.

Intergovernmental Panel on Climate Change (IPCC) (2014). *Climate Change 2014: Mitigation of Climate Change, Contribution of Working Group III to the Fifth Assessment Report of the Intergovernmental Panel on Climate Change*. Edenhofer, O., Pichs-Madruga, R., Sokona, Y., Farahani, E., Kadner, S., Seyboth, K., Adler, A., Baum, I., Brunner, S., Eickemeier, P., Kriemann, B., Savolainen, J., Schlömer, S., von Stechow, C., Zwickel, T. and Minx, J.C. (eds). Cambridge and New York: Cambridge University Press.

Intergovernmental Panel on Climate Change (IPCC) (2018). Summary for Policymakers. In *Global Warming of 1.5 °C. An IPCC Special Report on the Impacts of Global Warming of 1.5 °C above Pre-Industrial Levels and Related Global Greenhouse Gas Emission Pathways, in the Context of Strengthening the Global Response to the Threat of Climate Change, Sustainable Development, and Efforts to Eradicate Poverty*. Masson-Delmotte, V., Zhai, P., Pörtner, H.-O., Roberts, D., Skea, J., Shukla, P.R., Pirani, A., Moufouma-Okia, W., Péan, C., Pidcock, R., Connors, S., Matthews, J.B.R., Chen, Y., Zhou, X., Gomis, M.I., Lonnoy, E., Maycock, T., Tignor, M., and Waterfield, T. (eds). Geneva: World Meteorological Organization.

Keaney, M. (2014). Financialization and Social Structures of Accumulation Theory. *World Review of Political Economy*, 5(1), 45–77.

Klein, N. (2014). *This Changes Everything: Capitalism vs the Planet*. London: Allen Lane.

Malm, A. (2013). The Origins of Fossil Capital: From Water to Steam in the British Cotton Industry. *Historical Materialism*, 12(1), 15–68.

Malm, A. (2019). Against Hybridism: Why We Need to Distinguish between Nature and Society, Now More than Ever. *Historical Materialism*, 27(2), 156–187.

McIlroy, J. (2014). Marxism and the Trade Unions: The Bureaucracy versus the Rank and File Debate Revisited. *Critique: Journal of Socialist Theory*, 42(4), 497–526.

McIntyre, R. and Hillard, M. (2012). Capitalist Class Agency and the New Deal Order: Against the Notion of a Limited Capital-Labor Accord. *Review of Radical Political Economics*, 45(2), 129–148.

McMichael, P. (2000). Globalisation: Trend or Project? In R. Palan (ed.), *Global Political Economy: Contemporary Theories*. London: Routledge, 100–113.

Morton, A.D. (2007). *Unravelling Gramsci: Hegemony and Passive Revolution in the Global Political Economy*. London: Pluto Press.

Overbeek, H. (2000). Transnational Historical Materialism: Theories of Transnational Class Formation and World Order. In Palan, R. (ed.), *Global Political Economy: Contemporary Theories*. London: Routledge, 168–183.

Roper, B.S. (2013). *The History of Democracy: A Marxist Interpretation*. London: Pluto Press.

Rupert, M. (2005). Reading Gramsci in an Era of Globalising Capitalism. *Critical Review of International Social and Political Philosophy*, 8(4), 483–497.

Selwyn, B. (2015). Twenty-First-Century International Political Economy: A Class-Relational Perspective. *European Journal of International Relations*, 21(3), 513–537.

Sinclair, T.J. (2016). Robert W. Cox's Method of Historical Structures Redux. *Globalizations*, 13(5), 510–519.

Steffen, W., Broadgate, W., Deutsch, L., Gaffney, O. and Ludwig, C. (2015). The Trajectory of the Anthropocene: The Great Acceleration. *The Anthropocene Review*, 2(1), 81–98.

Thomas, P.D. (2010). *The Gramscian Moment: Philosophy, Hegemony and Marxism*. Chicago, IL: Haymarket Books.

Thomas, P.D. (2013). Hegemony, Passive Revolution and the Modern Prince. *Thesis Eleven*, 117(1), 20–39.

United Nations Environment Programme (UNEP) (2019). *Emissions Gap Report 2019*. Available at www.unenvironment.org/resources/emissions-gap-report-2019.

3 Competing ideas

Ecosocialist theory

Introduction

Ecosocialist contributions to the development and dissemination of radical climate justice ideas fall under the category of Cox's 'critical theory', and they compete with dominant 'problem-solving' approaches to the causes of, and proposed solutions to, the organic crisis of global capitalism. Ecosocialism is not a monolithic theoretical perspective, however, and there is much contestation in attempts to clarify the meaning of this concept. Theoretical clarification is important because, as prominent ecosocialist John Bellamy Foster emphasises, the understanding that theory ultimately informs practice is central to the Marxist tradition.

> Historically, Marxism has always taken the development of theory/ science very seriously, without which revolutionary praxis would be impossible. In the struggles to define the critique of capitalism embodied in Marxian ecology and ecosocialism it is essential to get the theory and the science correct to the extent possible. Our practice, the clarity of our ideas, our way forward depend on that.
>
> (Foster, cited in Stache, 2016)

The overview of 'varieties of ecosocialism' presented in this chapter aligns with Burkett and Foster's useful categorisation of first-stage, second-stage and third-stage ecosocialism which, following Holleman's (2015) caveat, are taken as 'shifts in the focus of debate' rather than linear developments within that body of thought referred to as 'ecosocialist' (Burkett, 2014a, 2014b; Foster, 2016; see also Stache, 2016). Given the extensive body of literature devoted to ecosocialist debates, the overview presented in this chapter is necessarily selective about what aspects of the debates are focused on while simultaneously trying to summarise enough information for the reader to understand the key issues under contestation. The overview of internal debates between first- and second-stage ecosocialists is followed by a brief summary of what Foster and other ecosocialists consider to be critically important theoretical debates between Marxist ecosocialist thinking and rival theories that support the pro-capitalist, technology-dense 'post-environmentalist' ecological modernisation project (as represented, for example, by groups such as the Breakthrough Institute). The chapter concludes with an overview of some of the work conducted by third-stage ecosocialists who apply Marxist ecological concepts and insights in their analyses of the current organic crisis of global capitalism. Third-stage ecosocialist contributions to the formulation of strategy and tactics by radical climate justice movement actors are discussed further in Chapter 6.

Competing ideas: varieties of ecosocialism

While many climate movement actors are becoming increasingly aware of the links between environmental, economic, political and social justice struggles, ecosocialists have developed and refined their understanding of these connections over many years, with 'second-stage' ecosocialists arguing that this understanding is evident in the original writings of Marx and Engels. Michael Löwy traces the origins of contemporary ecosocialist ideas to the 1970s, and states that the word 'ecosocialism' gained prominence within the German Green Party in the 1980s (Löwy, 2015, pp. xi–xii). As with all complex theoretical perspectives, however, there are many different interpretations of what 'ecosocialism' means. According to some theorists, ecosocialists combine *Marxist* critiques of capitalist political economy with *ecological* critiques of 'productivism' ('production as a goal in itself') and *wasteful* (or unnecessary) consumerism that serves profit accumulation rather than the purpose of meeting human needs in a sustainable way (Löwy, 2015; emphasis in original). When discussing production and consumption issues, for instance, many first-stage ecosocialists emphasise Marx's distinction between use value and exchange value when elaborating on their ideas about how ecologically sustainable consumption can be achieved (for example, refer to Kovel, 2007). According to some definitions, ecosocialism thus constitutes what has been called a 'red–green alliance' with the broad aim of building a new society with 'two fundamental and indivisible characteristics'.

> It will be *socialist*, committed to democracy, to radical egalitarianism, and to social justice … And it will be *based on the best ecological principles*, giving top priority to stopping anti-environmental practices, to restoring damaged ecosystems and to re-establishing agriculture and industry on ecologically sound principles.
>
> (Angus, 2011, p. 6; emphases in original)

While ecosocialists agree that ecological concerns would play a central role in the democratic and egalitarian socialist society they envisage and work towards creating, many of their disagreements revolve around the extent to which classical Marxism (as represented in the original works of Marx and Engels) needs to be modified to better incorporate an understanding of capitalism's environmental destruction. Debates amongst ecosocialist theorists are, in essence, disagreements about the extent to which ecosocialist theory needs to draw from 'green' perspectives (representing environmentalism) to supplement what some ecosocialists perceive to be ecologically deficient 'red' perspectives (representing socialism/Marxism). The differing analyses defended in these theoretical debates are important because of their implications for the future of the Marxist emancipatory project as originally conceived by Marx and Engels and, hence, for the strategies and tactics adopted by the labour movement and by environmental and climate justice actors. Thus, while some theorists argue that the variety of ecosocialist thought should be celebrated (for example, refer to Löwy, 2015, p. xv), 'second-stage' ecosocialists such as Foster and Burkett disagree. Critics of 'first-stage' ecosocialism argue that these debates create divisions that are potentially damaging to efforts to address the climate crisis because of their political and practical implications for decisions regarding agency, strategy and tactics.

First-stage ecosocialist theory

Despite the 'prefigurative Marxian environmental perspective' evident in the 1960s and 1970s work of authors such as Barry Commoner, Virginia Brodine and Howard Parsons (Foster, 2016, p. 395), the ecological content of Marx and Engels' work was largely neglected by most Marxists. Foster suggests a possible reason for this neglect with reference to Rosa Luxemburg's observation that Marx's vast body of work 'extended beyond the immediate needs of the working-class movement [and] would only be discovered and incorporated much later, as the socialist movement matured and new historical challenges arose' (Foster, 2015, p. 4). Thus one reason for the ecological content of Marx's work having been neglected by most Marxists in the past may have been because the ecological crisis had not yet unfolded to become as salient an issue as it is today. This lack of attention to what Marx wrote about capital's effects on the natural environment could have contributed to the widespread perception that Marx's writings were 'anti-ecological', leading to attempts to incorporate 'green theory' into Marxist theory. Foster describes this 'red–green alliance', which creates a 'hybrid analysis' of Marxism and environmental theory, as a feature of what he refers to as 'first-stage ecosocialism' (Stache, 2016).

Represented in the writings of theorists such as Ted Benton, André Gorz, James O'Connor, Joel Kovel and Daniel Tanuro, Foster argues that first-stage ecosocialist thinking 'developed under the hegemony of Green theory' (Stache, 2016) and was part of the self-criticism that began within Marxist theorising in the 1960s (Foster and Clark, 2016). Some of the major errors and omissions that first-stage ecosocialists identify in classical Marxist theory, which they argue must be amended to align with environmental perspectives, include its 'mechanistic-positivistic scientific view' of nature, its 'productivism' and promotion of 'Prometheanism' (human domination over nature) and the failure of the labour theory of value to incorporate environmental values. Foster argues that this perceived need to combine 'green' and 'red' theory, and to develop a new body of thought that negates and displaces 'classical socialism', is unnecessary because Marxism is *inherently* ecological (Stache, 2016). Other Marxists, such as Hannah Holleman, concur.

> [T]he ecological critique and imperative for change is already explicit in Marx's work and many socialist traditions. Therefore I consider myself a socialist rather than an ecosocialist, and the work described here is at heart simply Marxist in approach.
>
> (Holleman, 2015, p. 10, note 1)

Despite evidence that ecological concerns are incorporated in Marx and Engels' work, Foster (2016) identifies several possible reasons for the widespread perception during the 1980s and 1990s that this body of thought was 'anti-ecological'. These reasons include a misplaced blame of the environmental disasters of the Soviet Union under Stalin and subsequent Soviet administrations on 'Marxism', the general disarray within the Left after the fall of the Soviet Union, the ascendancy of postmodernist thinking that rejected Marxism in its entirety, a misinterpretation and oversimplification of Marx and Engels' views on nature as represented in their writings and a failure to understand that Marx uses the labour theory of value only for the purpose of critically analysing capitalism, not as a universal measure valid in all historical periods under all modes of production. In addition to the misunderstandings and misperceptions noted above, Foster suggests that in some cases the claim that Marx's work is 'anti-ecological' is ideologically motivated; one reason some:

> ecological scholars make such great efforts to ignore, downplay, or distance themselves from the [ecological] insights evident in Marx and Engels' work include Marxism's revolutionary potential in its focus on class struggle. Foster suggests that the idea of class struggle 'frightens' some Left academics and challenges both the conventional wisdom that 'the working class is by nature anti-environmental' and the normative political preferences of 'non-radical, capitalist-oriented environmentalists.'
>
> (Stache, 2016)

Critiques of first-stage ecosocialist theory

While it is extremely difficult (if not impossible) to address ideological motivations for rejecting arguments that classical Marxism is inherently anti-ecological, the *specific charges* made by first-stage ecosocialist theorists can be addressed by revisiting Marx and Engels' writings and pointing out the misinterpretations, omissions and misconceptions that have informed its perceived ecological shortcomings. In addition to conducting close textual analyses of Marx and Engels' original works to determine what they *actually* wrote and thought about ecology, second-stage ecosocialists also identify the origins of the widespread views that classical Marxism is 'anti-ecological' as this can help clarify some of the ideological motivations informing these views.

Foster and Clark (2016, pp. 3–6) attribute the origin of many of the misconceptions about Marx's views on ecology to critical theorist Alfred Schmidt's dissertation, which was published in Germany in 1962 as *Der Begriff der Natur in der Lehre von Karl Marx* (*The Concept of Nature in Marx*), and which, via several 'inconsistencies' and 'convolutions', misinterpreted Marx's work as positivist and mechanistic and claimed that he had promoted 'unrestrained productivism' (see also Burkett, 2005). Foster (2014, 2016) points out that Schmidt was a doctoral student working under the supervision of Frankfurt School critical theorists and Western Marxists Max Horkheimer and Theodor Adorno, who themselves had interpreted Marx's work as 'positivist'. The English translation of Schmidt's book was published in 1971, and its claim that Marx had 'fallen prey' to the Enlightenment project promoting 'Prometheanism' influenced the revived Marxism of the 1960s and 1970s, leading to the widespread view that Marx's thought does not incorporate concerns about the natural world (Foster, 2014, 2016). Foster argues that even studies of Marx's writings tend to ignore or gloss over the way in which his work incorporates 'comprehensive scientific analysis of the natural conditions underlying production and the capitalist economy' (Foster, 2014, p. viii).

Second-stage ecosocialist theory

The group of theorists that Foster refers to as 'second-stage ecosocialists' (a group that includes academics such as Paul Burkett, Brett Clark, Hannah Holleman, Stefano Longo, Kohei Saito, Richard York, and Foster himself) respond to first-stage ecosocialist critiques by revisiting Marx and Engels' original work in order to establish the accuracy of these claims. Second-stage ecosocialist research points to textual evidence that contradicts the conventional wisdom that classical Marxism ignores nature, or reduces its intrinsic and independent importance and treats it in purely instrumental terms. These researchers identify the centrality of capitalism's damaging impact on nature throughout Marx and Engels' work, pointing to many illustrative examples, and they argue that the way this body of work is read and interpreted by critics ignores this aspect of their writing.

Foster and Clark (2016) identify the emergence of 'second-stage' ecosocialism in the 1990s, with Peter Dickens' 1992 book *Society and Nature: Towards a Green Social Theory* representing an early example of the approach subsequently also adopted by other key second-stage ecosocialist theorists.[1] Dickens' approach constitutes a 'turning point' in ecosocialist theorising in that, rather than accepting the conventional wisdom that classical Marxism is anti–ecological and needs to be revised and supplemented with elements of 'deep-ecology' perspectives, he revisits Marx's early writings in order to analyse his theory on how capitalist relations of production alienate humanity from nature (Foster and Clark, 2016, p. 13). Dickens argues that instead of 'grafting' green theories onto a revised Marxism, it is necessary 'to extend Marx's method, which included both a historical-materialist and dialectical assessment of the relationship between society and nature' (Foster and Clark, 2016, p. 13).

Second-stage ecosocialists thus develop what Foster refers to as 'anti-critiques' in response to ecological critiques of Marxism by revisiting the original works by Marx and Engels with a specific purpose: 'to examine the role of ecological analysis in the deep structure' of their writings (Foster cited in Stache, 2016).[2] This re-reading of Marx and Engels found evidence that countered first-stage ecosocialist critiques so that over the next decade's debates, 'first-stage ecosocialists were forced to accede the ground at nearly every point' (*ibid.*). In the interview with Stache (2016), Foster considers the book *Marx and the Earth* (2016), which he co-authored with Burkett, as 'in many ways the culminating stage in this debate'.

> It is a response to a number of counterattacks and persistent misconceptions aimed at Marx and Engels, particularly in the area of ecological economics. Some ecological economists like Joan Martinez-Alier and James O'Connor argued that Marx and Engels failed to incorporate thermodynamics into their analysis. Similarly, it has been charged that Engels rejected the second law of thermodynamics. Other criticisms directed at classical historical materialism are also addressed, such as Joel Kovel's claim that Marx and Engels excluded any notion of the intrinsic value of nature, Daniel Tanuro's charge that Marx and Engels ignored the various qualitatively different forms of energy, and John Clark's contention that Marx denied the organic relations between nature and society.
>
> (Foster, cited in Stache, 2016)

In addition to responding to specific criticisms of Marx's work and questioning 'the tendency to pit the young Marx against the mature Marx, Marx against Engels, and natural science against social science' (Foster and Clark, 2016, p. 13), second-stage ecosocialists also focus on 'reconstructing' Marx and Engels' ecology (Foster, 2014). Close readings of Marx's writings reveal that nature plays a central role in his critique of capitalist political economy, particularly in his 'theory of metabolic rift, his ecological-value analysis, the analysis of ecological imperialism, and Marx and Engels' development of the dialectics of ecology' (Foster, cited in Stache, 2016; see also Foster, 2000, 2009). Marx's metabolic rift analysis, which is central to second- and third-stage ecosocialist analyses, is summarised later in this chapter.

Second-stage ecosocialist responses to specific critiques by first-stage ecosocialists

In addition to the ideological reasons that inform the widespread common perception that classical Marxism is anti-ecological (as noted previously), many confusions and misinterpretations of Marx's writings arise from a failure to recognise that the core of his work involves a *critique of capitalism* and not an ahistorical account of all social formations at all times. Thus, for example, Marx's labour theory of value constitutes an analysis of the form that value takes in *capitalist* societies, not 'of all labour at all times' (Foster and Clark, 2016, p. 9).

On the 'ecological deficiencies' of Marx's labour theory of value

One of the most significant and persistent critiques of classical Marxism made by mainstream green theorists as well as some first-stage ecosocialists is that the centrality of Marx's labour theory of value renders Marxist political economy antithetical to an ecological value analysis (Clark and Foster, 2010, p. 148). In response, Burkett observes that:

> Generally speaking, these critics fail to appreciate the historical and social-relational aspect of Marx's theory – that value as a specifically capitalist form of wealth does not represent Marx's normative valuation of nature's intrinsic worth (e.g. in terms of aesthetic and other use values).
> (Burkett, 2014a, p. 99)

Moreover, as Clark and Foster remind Marx's critics:

> The conceptual categories that Marx uses in his critique, such as nature as a free gift, and the law of value, are categories that he did not invent, but ones that he took over from classical political economy – *recognizing that they exhibited the real tendencies of the [capitalist] system [he critiques]* – and that he sought to transcend by transcending bourgeois society itself.
> (Clark and Foster, 2010, p. 149; emphasis added)

Burkett attributes many misinterpretations of Marx's value analysis to critics' failure to understand 'the distinctions and relations among Marx's conceptions of use value (to which nature always contributes), value (the necessary wage-labor time objectified in *commodity* use values), and exchange value (the monetary price paid for a use value)' (Burkett, 2014a, p. 69; emphasis in original). He goes on to explain the importance of these distinctions and relations to Marx's method in more detail.

> Only by clearly demarcating and showing the relations and tensions among value, exchange value, and use value phenomena is Marx able to

establish how capitalism's class-exploitative relations shape production together with its human and extra-human natural conditions. At the same time, Marx analyses how particular sub-forms of value and capital (e.g. money, wages, constant capital, fixed and circulating capital, rent) are themselves shaped by the material conditions of production, that is, by the natural basis and substance of use value. In this way, Marx's value analysis reveals the tensions between wealth in its capitalist form and wealth in the sense of the individual and collective needs of social human beings co-evolving with nature, along with the implications of these tensions for class struggle and the movement toward a new stage of wealth production.

(Burkett, 2014a, p. 100)

Importantly, Burkett goes even further in his response to critics who challenge Marx on his failure to ascribe an intrinsic value to nature, explaining that attempts to do this ultimately lead to conflations of different value forms that *obscure* the ecological contradictions inherent in capitalist economics.

[M]any of Marx's ecological critics want to directly attribute value to nature without taking account of the historical specificity of wealth's social forms as determined by particular production relations. As a result, when they try to specify the precise *value-form* taken on by nature (value in terms of what, and for whom?), they are driven to various theoretical contradictions and defaults. The most common contradiction here is the inability to define nature's purported 'value' independently of its exchange value and/or its use value, which often leads to (implicit or explicit) *conflations* of the three concepts. These conflations cause the critics in question to ignore or soft-pedal the ecological contradictions of capitalist wealth as revealed by Marx's relational and dialectical approach to value, exchange value, and use value.

(Burkett, 2014a, p. 100; emphasis in original)

Burkett (2014a, pp. 100–106) responds in detail to specific critiques of Marx's neglect of intrinsic value by Gunnar Skirbekk, David Orton, Geoffrey Carpenter and Ted Benton. These responses are very illuminating, and the interested reader is encouraged to consult Burkett's original writings. Although there are many additional intellectually interesting aspects to this debate (such as Marx's engagement with the Lauderdale Paradox, which Clark and Foster (2010) discuss in some detail), this summary of second-stage ecosocialist responses to critiques of Marx's labour theory of value concludes with Foster and Clark's reminder that:

it was … [the] very one-sidedness of the value form in capitalism that lay at the center of Marx's critique, associated with the contradiction between wealth (derived from natural-material use values) and value or exchange value (which left out nature altogether). For Marx, once it was recognized that nature – constituting, together with labor, one of the two sources of all wealth – was not included in the capitalist value calculus but was treated as a 'free gift ... to capital,' it was impossible *not* to recognize both the existence of natural limits and capital's destructive tendency to override them, in its unending drive for accumulation.

(Foster and Clark, 2016, p. 9)

On Marx's 'productivism' and views on the human conquest of nature

Related to critiques about Marx's purported neglect of nature's intrinsic value in his value theory are the claims that classical Marxism is anti-ecological because of its 'productivism' and 'Prometheanism'. While Burkett addresses such claims in greater detail in his 2014 book *Marx and Nature: A Red and Green Perspective*, this is also the main theme of his paper entitled 'Marx's Vision of Sustainable Human Development' (Burkett, 2005). In this paper, Burkett sets out to address critiques that Marx's vision of 'communism or socialism (two terms that he used interchangeably)' implies an 'ecologically unsustainable' assumption that there are no natural limits to production and that Marx is 'Promethean' in that he supported and promoted the Enlightenment project of achieving human domination over nature.

Burkett begins his discussion by identifying a few examples of such critiques, including one made by ecological economist Herman Daly, who contends that for the 'materialist determinist' Marx, 'economic growth is crucial in order to provide the overwhelming material abundance that is the objective condition for the emergence of the new socialist man and that environmental limits on growth would contradict "historical necessity"' (Daly, cited in Burkett, 2005, pp. 34–35).[3] Burkett (2005, p. 35) also refers to Robyn Eckersley's claim that 'Marx fully endorsed the "civilizing" and technical accomplishments of the capitalist forces of production and thoroughly absorbed the Victorian faith in scientific and technological progress as the means by which humans could outsmart and conquer nature' and her argument that Marx 'consistently saw human freedom as inversely related to humanity's dependence on nature'. Briefly outlining a few other critiques along similar lines, Burkett points out the political implications of accepting such critiques without engaging with Marxism's 'human developmental and ecological elements', which are that they lead even some Marxists to promote the '"greening" of capitalism as a practical alternative to the struggle for socialism' (*ibid.*).

Given the political and practical implications of this debate, Burkett returns to the writings of Marx and Engels to identify the key elements of their vision of 'sustainable human development' within a post-capitalist economy and society in order to examine whether these visions were, in fact, 'anti-ecological' or 'productivist', and whether they promoted (or supported) 'the human domination of nature'.[4] He points out that the 'conventional wisdom' that Marx and Engels rejected speculations about 'socialist utopias' and provided few thoughts about what a post-capitalist society should look like ignore the fact that 'post-capitalist economic and political relationships' are a recurring theme in all their major works and most of their minor works (Burkett, 2005, p. 36). This theme is also explored in Peter Hudis' book, *Marx's Concept of the Alternative to Capitalism* (2012), which provides a detailed scholarly investigation of Marx and Engels' vision of a desirable post-capitalist future. Despite this vision of a post-capitalist future being 'scattered' throughout the entire body of their work, Burkett argues that 'one can easily glean from them a coherent vision based on a clear set of organizing principles', which he describes as follows.

> The most basic feature of communism in Marx's projection is its overcoming of capitalism's social separation of the producers from necessary conditions of production. This new social union entails a complete decommodification of labor power plus a new set of communal property rights. Communist or 'associated' production is planned and carried out by the producers and communities themselves, without the class-based intermediaries of wage-labor, market, and state. Marx often motivates and illustrates these basic features in terms of the primary means and end of associated production: free human development.
>
> (Burkett, 2005, p. 36)

More specifically, Burkett argues that three aspects of Marx's vision of a desirable post-capitalist future should be considered when evaluating the extent to which this vision is environmentally sustainable: '(1) the responsibility of communism to manage its use of natural conditions; (2) the ecological significance of expanded free time; [and] (3) the growth of wealth and the use of labor time as a measure of the cost of production' (Burkett, 2005, p. 45). He notes that not only does Marx's writing clearly demonstrate his deep concerns about the way in which capitalist agricultural practices saps 'the original sources of all wealth, the soil and the labourer', but he also 'repeatedly emphasizes the imperative for post-capitalist society to manage its use of natural conditions responsibly' and insists 'on the extension of communal property to the land and other "sources of life"' (Burkett, 2005, p. 46). Communal land ownership, in Marx's vision, comes with the responsibility to treat 'the soil as *eternal* communal property, an *inalienable* condition for the existence and reproduction of a chain of successive generations of the human race' (Marx, cited in Burkett, 2005, p. 47). Instead of treating 'people' and 'nature' as 'two separate "things"', Marx and Engels consider humanity to have 'an historical nature and a natural history' (*ibid.*). In addition, they regard individuals as 'subservient to nature', and communism as a system that would overcome ruptures between humans and nature, with Marx going 'so far as to define communism as "the unity of being of man with nature"' (*ibid.*).

Given the material reality that humans cannot survive unless they meet their needs by interacting with nature, in Marx's vision of a communist future this interaction will, of course, continue – but it will be conducted by 'the associated producers *rationally regulating* their interchange with nature' (emphasis added). This rational regulation presumes that the producers control 'their own social organisation', not that 'natural limits' no longer exist (Burkett, 2005, p. 47). Marx's understanding that humans cannot 'master' nature is evident in the suggestion that 'a portion of the surplus product' should be set aside 'to provide against misadventures, disturbances through natural events, etc.' (Marx, cited in Burkett, 2005, p. 48). Burkett summarises Marx and Engels' views about the ability of humans to control nature as follows.

> Contradicting their ecological critics, Marx and Engels simply do not identify free human development with a one-sided human domination or control of nature. According to Engels, 'Freedom does not consist in the dream of independence of natural laws, but in the knowledge of these laws, and in the possibility this gives of systematically making them work towards definite ends' … In short, Marx and Engels envision a 'real human freedom' based on 'an existence in harmony with the established laws of nature.'
>
> (Burkett, 2005, pp. 48–49)

Having addressed the claims of critics that Marxism is inherently anti-ecological because of its alleged promotion of 'human domination over nature', Burkett turns his attention to arguments that the Marxist vision of 'expanded free time' is inherently anti-ecological. Such arguments are based on assumptions that labour can only enjoy more free time if there is extensive automation (using energy-intensive technology) or that there are such abundant resources that people do not need to work long hours (which implies the availability of limitless natural resources). Burkett (2005, pp. 49–50) points out that 'the ecological critics have mischaracterised the relation between free time and work time under communism', a system in which not only is producers' labour time 'reduced to a *normal length*' (emphasis added) because they are no longer compelled to work for others, but the labour itself is also qualitatively different and presents opportunities for the achievement of higher self-realisation. Burkett summarises the ecological nature of Marx and Engels' vision of free time as follows.

> In short, the founders of Marxism did not envision communism's reduced work time in terms of a progressive separation of human development from nature. Nor did they see expanded free time being filled by orgies of consumption for consumption's sake. Rather, reduced work time is viewed as a necessary condition for the intellectual development of social individuals capable of mastering the scientifically developed

forces of nature and social labor in environmentally *and* humanly rational fashion. The 'increase in free time' appears here as 'time for the full development of the individual' capable of 'the grasping of his own history as a *process*, and the recognition of nature (equally present as practical power over nature) as his real body.' ... Far from anti-ecological, this process is such that the producers and their communities become more theoretically and practically aware of natural wealth as an eternal condition of production, free time, and human life itself.

(Burkett, 2005, p. 50; emphasis in original)

Burkett (2005, p. 51) furthermore points out how 'ecological critics also seem to have missed the potential for increased free time as a means of *reducing* the pressure of production on the natural environment' (emphasis in original), which can happen in a system in which rising productivity is rewarded by free time rather than by more money with which to buy more consumer goods. Decreased consumption is furthermore linked to 'communism's transformation of human needs', which constitutes the next part of Burkett's argument against critics who charge Marxism with being anti-ecological.

Addressing the issue of Marx and Engels' 'notorious references to continued growth in the production of wealth under communism', Burkett (2005, p. 52) emphasises that rather than being 'productivist', Marx and Engels define 'growth' in terms of 'free and well-rounded human development', not 'material production and consumption for their own sake'. He points out that when Marx and Engels talk about growth, therefore, it is always with reference 'to growth of wealth in a general sense, encompassing the satisfaction of needs other than those requiring the industrial processing of natural resources (matter and energy throughput)' (*ibid.*). Citing Ernest Mandel, Burkett (2005, p. 53) points out that the Marxist 'social and human developmental approach to need satisfaction is quite different from the "absurd notion" of unqualified "abundance" often ascribed to Marx'. Needs can be categorised as 'basic', 'secondary' and 'luxury, inessential or even harmful needs', and, drawing on Mandel's work, Burkett (2005, p. 53) argues that Marx foresees communist societies as meeting everyone's *basic* needs followed by 'the gradual extension of this satisfaction to secondary needs ... *not* a full satiation of all conceivable needs' (emphasis in original). He furthermore argues that it is in Marx's notion of secondary needs that 'one begins to see the full ecological significance of free time as a measure of communist wealth', particularly 'if the secondary needs developed and satisfied during free time are less material and energy intensive'. Burkett points out that this is, indeed, precisely Marx's vision; he sees 'the producers using their newfound material security and expanded free time to engage in a variety of intellectual and aesthetic forms of self-development' (*ibid.*).[5] Foster and Clark (2016, p. 4) also conclude from their readings of Marx's work that, rather than being a 'productivist, he privileged the goal of the fulfilment of humanity's *qualitative* needs (use value) ... He saw value, which in capitalist economics emanates from "capital alone," as "contradicting" wealth'. The two fundamental sources of wealth in Marx's analysis, nature and labour, are relevant in another important debate between first-stage and second-stage ecosocialists over the validity of James O'Connor's thesis of the 'second contradiction of capitalism'.

On Marx's failure to identify the 'second contradiction of capitalism'

Given that nature is an essential source of wealth, and that 'free nature' has been pivotal to accumulation strategies throughout the history of capitalism, first-stage ecosocialist James O'Connor considers how the serious environmental degradation that was already evident in the 1980s (and is much more serious now) might negatively impact on capital accumulation. O'Connor proposes a modification to Marx's theory: the incorporation of a 'second contradiction of capitalism' which, he argues, like Marx's 'first contradiction of capitalism', could also lead to *economic* crises. O'Connor introduced his thesis of the 'second contradiction of capitalism' in a paper published in the 1988 inaugural volume of a leading academic ecosocialist journal, *Capitalism Nature Socialism*, and expanded on it in later writings (Foster, 1992, p. 77). In this paper, O'Connor argues that while identifying 'the first contradiction of capitalism', which (according to O'Connor's interpretation of Marx) is its 'inherent tendency toward a realization crisis, or crisis of capital over-production', Marx fails to identify 'the second contradiction of capitalism': the degraded 'conditions of production' it generates as it depletes natural resources and pollutes and destroys ecosystems (Foster, 1992, pp. 78–79). O'Connor furthermore argues that the post-1970s modifications to global capitalism have enabled an 'intensified exploitation (and superexploitation) of labor and the environment' which is so severe that the 'second contradiction of capitalism' is 'rapidly gaining on the first' and becoming the deciding contradiction 'marking nature's ultimate "revenge" on the accumulation process' (Foster, 1992, p. 81). While Marx recognised that capitalist agriculture was environmentally damaging, O'Connor argues that he failed to consider the possibility that the ecological degradation of the capitalist mode of production might itself lead to an 'economic crisis of a particular type, namely, underproduction of capital' (O'Connor, cited in Foster, 2002, p. 7). According to O'Connor, this failure on Marx's part requires ecological Marxists to develop 'a theory of how increasing ecological costs [contribute] to decreasing profitability and accumulation crisis' (Foster, 2002, p. 7). Before discussing Foster's response to O'Connor's argument regarding the overriding importance of 'the second contradiction of capitalism', it is pertinent to comment on the differences between O'Connor's thesis and Marx's understanding of capitalism's *fundamental* contradiction.

Contrary to popular perceptions that Marx located capitalism's 'historical limits' solely to 'tendencies toward overaccumulation and falling profitability', Burkett points out that Marx identifies capitalism's 'fundamental contradiction' as 'the contradiction between production for private profit and production for human needs' (Burkett, 2014a, pp. 175, 177). Thus, in contrast to O'Connor's understanding of what he refers to as the 'first contradiction of capitalism', Burkett points out that, 'For Marx, capitalism's fundamental contradiction is not reducible to accumulation crises; rather, such crises "reveal" this fundamental contradiction' (Burkett, 2014a, p. 180). Moreover, Burkett argues that

> the conflict between production for profit and production for human needs, the alienation of the conditions of production vis-à-vis the pro-ducers and their communities, and the tension between social production and private appropriation, are all equivalent expressions of capitalism's fundamental contradiction in Marx's view.
>
> (Burkett, 2014a, p. 178)

O'Connor's 'two contradictions' are therefore, from a Marxist perspective, 'both *symptoms* of this more basic contradiction' (*ibid.*, p. 196; emphasis in original). Burkett summarises his critique of O'Connor's thesis as follows.

> For Marx, the fundamental contradiction of capitalism is that between wealth *for capital* versus wealth *for the producers and their communities* – where the latter is defined not in terms of the minimalist material and social requirements of capital accumulation but rather in terms of the conditions for a less restricted and more sustainable human development … Marx does not artificially divide capital's power over both labor and its conditions into two separate powers. Rising exploitation, overproduction crises, increasing 'external costs' of production, and the degradation of human, natural, and social wealth are all necessary, mutually constituted aspects of capitalism's fundamental contradiction, in Marx's view.
>
> (Burkett, 2014a, p. 197; emphases in original)

Supplementing Burkett's approach of returning to Marx's original writings to review his theory of the fundamental contradiction of capitalism, Foster (2002, p. 6) considers two issues in evaluating O'Connor's argument: whether ecological crises necessarily lead to economic crises under capitalism, and the extent to which ecological contradictions lie 'at the heart of capitalist society'.

Regarding the issue of whether ecological crises necessarily lead to capital-ist economic crises, Foster (2002, pp. 10–11) argues that it would be a mistake to 'underestimate capitalism's capacity to accumulate in the midst of the most blatant ecological destruction, to profit from environmental degradation … and to continue to destroy the earth to the point of no return – both for human society and for most of the world's living species'. Burkett (2014a, p. 195) similarly argues that the problems created by environmental degradation can present new opportunities for realising surplus value, and empirical evidence supports this view. Naomi Klein's (2007, 2017) works on 'disaster capitalism,' as well as other analyses inspired by this work, provide examples of how capitalists already profit from environmental degradation and disasters and also outline some of the neoliberalising commodification and financialisation mechanisms and processes being used to profit from 'natural resource management', neoliberal forms of nature conservation and environmental disasters such as the climate change caused by anthropogenic global warming.

Fletcher (2012), for example, discusses several ways in which the need to address climate change is being used as a justification for financialising nature conservation through projects such as Reducing Emissions from Deforestation and Forest Degradation (REDD+) and the Clean Development Mechanism (CDM) as well as through the creation of carbon trading markets and environmental (or 'weather') derivatives. Extending the profit opportunities provided to capital by financial instruments facilitating speculation on staple food crops that result in price hikes disadvantaging the poorest people (Baines, 2017; Clapp, 2009), these additional financial instruments have been created to allow speculators to gamble on the weather (Pike and Pollard, 2010; Randalls, 2010).[6] These derivatives include 'catastrophe bonds', which are designed to 'manage the risks of improbable but catastrophic natural events' (Fletcher, 2012, p. 107).

Randalls (2010, p. 711) defines weather derivatives as 'financial contracts that enable companies to trade upon weather indices (such as temperature, precipitation, snowfall, wind velocity or frost) to manage their weather-sensitive costs or simply to speculate'. He identifies the emergence of these derivatives to 'the US energy sector in the mid-1990s with Enron, Aquila and Koch Industries' (*ibid.*). 'Weather derivatives' enable wealthy financial speculators, whose profits and lifestyles are largely to blame for anthropogenic global warming (AGW), to gamble on and profit from the resulting climate change crisis while the poor and disadvantaged people whose contribution to AGW is negligible suffer, many even losing their lives as a result of 'extreme weather' events manifesting as prolonged droughts and extreme floods. Fletcher concludes that:

> [i]n this way, uncertainty concerning climate change impacts becomes not a hindrance to marketization but yet another opportunity for profit; both the climate crisis and uncertainty concerning the same become distinct sources of value, a double reversal of James O'Connor's (1994) predictions.
>
> (Fletcher, 2012, p. 107)

Climate movement actors such as Naomi Klein argue that these opportunities to profit from this model of 'disaster capitalism' may even motivate efforts to ignore or, worse still, actively 'discredit predictions of the impending climate crisis, in order to harness both current sources of profit potentially compromised by a serious mitigation response and, moreover, to let the crisis unfold in anticipation of the new sources of profit thereby created', as evidenced by how ExxonMobil funded climate change denial (Fletcher, 2012, pp. 107–108).

In addition to the ways in which capital can (at least in the short term) transform environmental crises into profit-making opportunities, Foster raises two more issues that O'Connor fails to address when formulating his thesis on the 'second contradiction of capitalism': that environmental damage could be serious while not directly affecting conditions of production, and the implications of the fact that effects of environmental damage are not the same everywhere. Listing several examples, such as the serious degradation of the ozone layer and the extinction of species that are 'still unknown to science', Foster (2002, p. 11) points out that the most severe environmental damage of the normal operations of the global capitalist economy does not necessarily occur 'where it principally affects the conditions of production' and it thus need not result in accumulation crises. With respect to the implications of the uneven effects of the damage caused by environmental disasters such as anthropogenic global warming leading to climate change, Foster points to the example of the Bush Administration's *Climate Action Report, 2002*, in which:

> The EPA acknowledged the dangers to life and living conditions represented by global warming, but emphasised that in the United States the environmental damage would be most visible in the melting of snow in the mountains, and the like. Where the conditions of production of agriculture were concerned, global warming, it was suggested, might even increase overall agricultural productivity. This lack of a clear connection between environmental damage and damage to the economic conditions of production was used (via standard cost-benefit analysis) to justify a policy of *adapting* to global warming as it developed, rather than taking measures to decrease the extent of global warming – since these would increase the costs of production. It follows that there is no natural feedback mechanism that automatically turns environmental destruction into increasing costs for capital itself.
>
> (Foster, 2002, p. 12; emphasis in original)

Foster concludes that a focus on capitalism's undermining of its own conditions of production therefore 'downplay[s] the full dimensions of the ecological crisis and even of capitalism's impact on the environment in the process of trying to force everything into the locked box of a specific economic crisis theory'. Like Burkett, who points out that the effect of O'Connor's dichotomy 'tends to soften the distinction between the conditions required for capitalist production and the conditions required for human development' and leads to artificial divisions between 'labor and ecological struggles – with the latter still basically defined as "non-class"' (Burkett, 2014a, p. 197), Foster similarly identifies the important political implications of O'Connor's argument: the 'second contradiction of capitalism' is 'tied to the growth of contemporary radical social movements' while the 'first contradiction of capitalism' (as O'Connor interprets it) 'is associated with the class-based labor movement' (Foster, 2002, p. 9). O'Connor's hope that social movements will succeed in forcing capital to internalise its current 'externalities' is also unlikely to be realised; what is more likely is that environmental destruction will instead 'provide entirely new ways to profit' (Foster, 2002, p. 12) which, as discussed above, is precisely what is happening.

An even more important issue to consider in the context of the climate change crisis is the danger that capital's ability to transform environmental crises into profit-making opportunities in the short term does not mean that the market mechanisms deployed to profit from these new accumulation opportunities will also do what they claim to do: address the ecological crisis effectively. As Fletcher notes:

> research is needed to investigate the key question raised by this analysis: To what extent does all of this actually contribute to effectively mitigating the climate change impacts it purports to address? After all, critics question whether carbon markets truly effect a net emissions reduction or merely conceal continued carbon production through sleight-of-hand accounting.
>
> (Fletcher, 2012, p. 108)

Fletcher extends the example of the failure of carbon trading to decrease GHG emissions to how other projects (such as hydroelectric dams), while being promoted as decreasing GHG emissions, actually release methane that results in even larger CO_2-equivalent GHG emissions. These examples lead to the argument that:

> [d]ynamics such as this demand further investigation in order to assess the extent to which the swiftly growing campaign to address climate change through neoliberal carbon market mechanisms is in fact capable of contributing to an effective resolution of the impending crisis rather than merely stimulating capitalist expansion.
>
> (Fletcher, 2012, p. 109)

Empirical research of this nature is also necessary if one is to more effectively counter arguments made by proponents of ecological modernisation (or 'eco-modernism') and the theorists whose work is used to support their project, such as those associated with 'production of nature', 'radical social constructionist hybridity' and 'radical social monist' perspectives.

Competing ideas: ecosocialism versus ecomodernism

Promoting technological innovations and the extension and intensification of capitalist relations of production as the best ways to address contemporary environmental crises, ecological modernisation theory stands in direct opposition to ecosocialism. The brand of 'post-environmentalist' ecological modernisation founded by Ted Nordhaus and Michael Shellenberger, which is the focus of debates discussed here, commenced from their critique of the major US environmental organisations as espoused in a 2004 'influential and controversial' pamphlet, *The Death of Environmentalism: Global Warming Politics in a Post-Environmental World* (Buck, 2013). While many environmentalists criticised the major ENGOs for their complicity with the US government's failure to legislate strong environmental measures, Nordhaus and Shellenberger's critique of environmentalism takes a very particular form – it criticises US environmentalists for lacking a large 'patriotic' vision to help them build alliances with government, business and civil society in order to support economic growth and develop the technologies they see as necessary for solving environmental crises.[7] The ecomodernist 'post-environmentalist' label signals this perspective's affinity with postmodern and 'production-of-nature' theoretical perspectives, which Foster argues serve the purpose of sowing confusion about the causes of environmental degradation in order to gain credibility for their project and win over supporters from the wider Left.[8]

Foster identifies the 'production-of-nature' perspective associated with intellectuals such as Neil Smith and Noel Castree as a separate 'influential tradition' of environmental thinking that developed in the 1980s and 1990s within the discipline area of 'radical geography' (Foster, 2016, p. 396). Replacing Schmidt's 'negative critique of the domination of nature' with what was claimed to be a 'more positive view of the production of nature' ultimately led to what Foster refers to as 'a left social constructionism and social monism, merged with political-economic perspectives, in which *nature was seen as subsumed within society*' (*ibid.*, emphasis added). Overlapping with Bruno Latour's emphasis on the '"hybridity" of society and nature', and drawing on long-standing philosophical debates over Cartesian dualism, radical social constructionist and social monist theorists argue that, despite his dialectical perspective, Marx 'fell prey' to the 'nature–society dualism' and that Marxism is therefore 'fatally flawed' because its founders 'failed to perceive the emergence of a hybrid world … populated by networks of machines, artifacts, cyborgs, etc. or as Latour says "monsters"' (Foster, 2016, pp. 396–397). The political implications of this perspective are evident in how these ideas are used to justify the ecomodernist project (Foster, 2016, p. 398). In a talk Latour gave at the Breakthrough Institute, for example, he said that 'the object today should be to "Love Your Monsters" (2012),' an idea Foster elaborates on as follows.

> In this view, 'imbroglios' or 'technological monsters', modern versions of Mary Shelley's Frankenstein, are a normal part of our relation to nature, and we should accept them and their consequences, while rejecting environmentalism in favour of 'political ecology' that consciously internalizes or bundles nature.
>
> (Foster, 2016, p. 398)

Foster points out that Latour's views align with the thinking of Nordhaus and Shellenberger, who do not accept any notion of natural limits to capital accumulation and unlimited growth but propose a 'breakthrough', by which they mean a 'post-environmentalism' that emphasises technological and market-based solutions to environmental crises. It should be noted, however, that although Foster does not discuss this, Bruno Latour joined the chorus of other critics of the Breakthrough Institute's ecomodernist project who published their critiques in Volume 7 of the journal *Environmental Humanities* (2015). Critiques of the Breakthrough Institute's ecomodernist theorists in that journal include the way in which it presents simplistic arguments that are not empirically supported while simultaneously ignoring existing empirical evidence that contradicts its arguments, and the way in which its arguments are incoherent and sometimes even inconsistent with its own stated goals of 'saving the environment' (Buck, 2013; Hamilton, 2015, 2017; Latour, 2015; Monbiot, 2015; Szerszynski, 2015). In his contribution to this collection of critiques, Latour expresses his misgivings about 'this monster, "ecomodernism," that I am not sure we should learn to love, and that triggers in me, I have to confess, a deep antipathy' (Latour, 2015, p. 220). One of the points Latour emphasises in this paper is the failure of the ecomodernists to understand the significance of the Anthropocene, a point that Hamilton focuses on in his critique of the Breakthrough Institute's vision of ecomodernism.

Hamilton, professor of public ethics and well-known author of many books trying to alert the public to the dangers of large scale geo-engineering projects as 'solutions' to AGW (see, for example, Hamilton, 2013), presents a particularly scathing assessment of ecomodernists who refer to a '*good* Anthropocene' (Hamilton, 2017, pp. 22–23; emphasis in original). Hamilton emphasises that the notion of a 'good Anthropocene' is an oxymoron that is both inappropriate and dangerous; it is used by those who welcome the opportunities that a disturbed Earth System presents 'to prove our ingenuity and technological facility' (*ibid.*). Hamilton's critiques of these ecomodernist arguments succinctly describe both the philosophy informing this project as well as the dangers it poses.

> For the ecomodernists, instead of final proof of the dangers of hubris, the new epoch is greeted as a sign of humankind's ability to renovate and control nature … In this eco-Promethean view, the Anthropocene is not evidence of human short-sightedness or foolishness, nor of global capitalism's rapaciousness, but presents an opportunity for humans finally to come into their own … For the ecomodernists, if we are capable of developing technologies to control the climate and regulate the Earth as a whole, then why not? Planetary engineering reframes global warming. No longer a vindication of environmentalist warnings that humans have gone too far, climate change becomes the spur to final victory for the human mastery project.
>
> (Hamilton, 2017, pp. 23–24)

Hamilton argues that not only are the technological fixes ecomodernists propose for a 'good' Anthropocene unlikely to work, but they also constitute a continuation along a course that will exacerbate unpredictable and dangerous shifts in the Earth System (see Hamilton, 2017, pp. 70–71). In addition to identifying these flaws in ecomodernist thinking, Hamilton also critically analyses several other theorists' positions on the disruption of the Earth System's functioning, including Jason Moore's thesis that the origins of this outcome lie in the 'Nature/Society binary' that lies at the heart of the modernist project. Hamilton argues that Moore's epistemology, which 'blurs the distinction between scientific facts and social facts', leads to an 'impossible contradiction', and that such thinking is not uncommon in contemporary social science disciplines.

> That Moore cannot distinguish between geological history and human history is symptomatic of much contemporary critical social science. He takes to its extreme the argument that we must dissolve all Cartesian dualisms, that is, the divide between nature and culture, pursuing an ontological flat-land of entanglement … Transcending science altogether, Moore ends up rejecting the claim that we are living in the Anthropocene because it is 'a curiously Eurocentric vista of humanity.' And the determination to reject all dualisms sees him challenging the foundational scientific claim that 'humans are overwhelming the great forces of nature.' We cannot overwhelm nature when we are indistinguishable from it. How could social science come to this? The impossible contradiction in Moore's position now becomes clear. On the one hand, he wants to deny humans their power and special place with a 'post-humanist' embedding of humans in nature; on the other hand, he wants to define the new epoch in terms of historical relations of human power and exploitation.
>
> (Hamilton, 2017, pp. 96–97)

Hamilton's critiques of ecomodernist thinking touch on some of the core issue that concerns ecosocialists about this philosophy: the way in which aspects of postmodern thinking open it to appropriation as a tool by those who seek to spread confusion and thereby remove attention from both the causes and the severity of the environmental damage caused by the Great Acceleration. As Foster (2016, p. 401) points out, Moore deploys his rejection of 'Cartesian binaries' to critique ecosocialist theorists who use Marx's concepts of the universal metabolism of nature, social metabolism and the metabolic rift in their dialectical analyses critiquing capitalism's environmental destruction. In opposition to the 'binaries' of 'nature' and 'society', Moore:

> substitutes his own 'singular metabolism,' which is nothing other than the idealized capitalist notion of the market expanded to encompass the entire web of life. This view adamantly rejects the whole notion of 'natural limits,' or the idea that in numerous cases ecological 'limits are

outside of us' … constituting insuperable barriers to production … To point to antagonistic relations between capitalism and nature (or to conceive of nature as apart from society even by means of abstraction) is for Moore … to fall prey to the 'Cartesian divide.'

(Foster, 2016, p. 405)

In an interview with Ian Angus, the transcript of which was published on the ecosocialist website Angus founded and edits, Climate & Capitalism (climate andcapitalism.com), Foster concludes that Moore may have set off to 'update or deepen Marxism' but has, instead, 'ended up by abandoning Marxism's revolutionary essence and adapting to capitalist ideologies' (Angus and Foster, 2016). The rejection of historical materialist dialectics and critical realism on the grounds that this analytical framework constitutes a simplistic 'Cartesian dualism' ultimately leads to a social monism that is incapable of understanding the 'complex mediations between nature and society within a dialectical concept of totality' (Foster, 2016, p. 399). Even more problematically, Foster points out that 'for many social constructionists, radical postmodernists, and left idealists, the problem of nature is essentially eliminated through its subordination to society', as is natural science itself, 'since natural processes are now to be treated as internal to the social dialectic' (*ibid.*, pp. 399–401). In contrast to radical social monist positions, Marx and Engels' historical materialism provides a powerful analytical tool that accounts for and accommodates 'dynamics, complexity, contradiction, emergence, and transformation in the analysis of the world at large' (Foster, 2016, p. 414). Historical materialism enables Marx to analyse the way in which human labour, which is a necessary activity that mediates between humans and nature in all modes of production, causes metabolic rifts in natural biophysical cycles and processes in the context of capitalist relations of commodity production (*ibid.*).[9] Second-stage ecosocialists highlight the relevance of Marx's metabolic rift analysis to explaining the way in which late capitalism causes ecological rifts on a planetary scale.

Marx's ecology: metabolic rift analysis

Referring to a range of his writings, Foster and Clark (2016) point out that Marx incorporated German chemist Justus von Liebig's concept of 'metabolism' in his critique of political economy.[10] Marx refers to the biophysical cycles and processes that 'constitute and help regenerate ecological conditions' as the 'universal metabolism of nature', an 'earthly metabolism' within which humans exist as they continually interact with the natural environment to meet their needs and produce goods and services, and he refers to the *labor process* (which includes 'exchanges with ecological systems') as the 'social metabolism' (Foster and Clark, 2016, p. 15). Marx argues that different modes of production generate 'distinct social metabolic orders' and, unlike in previous socio-ecological systems, the social metabolism of capitalist commodity production 'generates ecological crises, manifesting as a "rift" in the metabolism between society and nature (or disjunctions within both the social metabolism and the wider universal metabolism)' (*ibid.*). In the capitalist mode of production, which is defined by a 'compulsion to accumulate', capital's needs 'are imposed on nature, increasing the demands placed on ecological systems and the production of wastes' and creating what Marx refers to as a 'metabolic rift' (*ibid.*, p. 16).

Marx's insights about the metabolic rift that results from capitalist commodity production and its spatial organisation are evident in his writings about the most serious ecological issue attracting the attention of scientists in his lifetime: the issue of soil fertility. Providing a historical account of how soils retained their fertility in pre-capitalist modes of agricultural activity by recycling both animal and human waste in the land where food was grown because it was consumed at the same place or nearby, Marx analyses how capitalism transformed this particular metabolic interchange with reference to the enclosure movement that forced peasants off the land, alienating them from their means of production so that they had to seek employment in the industrial centres located in towns. The greater separation between town and country that the new industrial systems necessitated also led to the transfer of soil nutrients from one location to another 'as food and fibre from farms were increasingly shipped to distant markets' (Foster and Clark, 2016, p. 15). The expansion of capitalist relations of production thus alienated the producers from nature and simultaneously led to both the 'squandering' of nutrients and increased levels of pollution in the towns and rivers that had to absorb the wastes (*ibid.*). In addition, Marx also discusses the way in which profit-maximising capitalist agricultural production practices 'increased the scale of operations, transforming and intensifying the social metabolism while exacerbating the depletion of the soil nutrients' (Foster and Clark, 2016, p. 15). Citing Marx (*Capital*, Volume 1), Foster and Clark (2016, p. 17) draw attention to his argument that capitalism thus creates 'a metabolic "rift" in the soil nutrient cycle, "robbing the soil" and "ruining the more long-lasting sources of that fertility"'. They identify the metabolic rift theme as one that recurs throughout Marx's writing; for example, in *Capital*, Volume 3, Marx argues that 'the drive to capital accumulation':

> reduces the agricultural population to an ever decreasing minimum and confronts it with an ever growing industrial population crammed together in large towns; in this way it produces conditions that provoke an irreparable rift in the interdependent process of social metabolism, a metabolism prescribed by the natural laws of life itself. The result of this is a squandering of the vitality of the soil, which is carried by trade far beyond the bounds of a single country.
> (Marx, cited in Foster and Clark, 2016, p. 17)

Marx also writes about how attempts to compensate for the loss of soil fertility in England by importing 'millions of tons of guano and nitrates from

Peru and Chile' constitute a form of 'ecological imperialism', contributing to a 'global metabolic rift', and he notes how such 'artificial solutions' compound the overall environmental degradation that is a by-product of intensive agricultural production (Clark and Foster, 2010, p. 146). Marx's analysis of the nineteenth-century ecological crisis of soil depletion constitutes a valuable illustrative case study of how to apply the method of what Foster and Clark (2016) refer to as 'ecological materialism' to analyse contemporary ecological issues, as many third-stage ecosocialists have done. Moreover, Foster and Clark (2016, p. 22) argue that the 'enduring value of Marx's ecological materialism' is how it highlights 'the need for a new order of social metabolic reproduction rooted in substantive equality'. Expanding on this point, they state:

> [h]ere social and natural necessity, natural science and social science, humanity and the earth become one human-mediated totality, in a wider universal struggle – one pointing to a revolutionary dialectic of humanity and the earth in which the necessary outcome is a world of sustainable human development. It is this higher synthesis of the various Marxian ecological and social critiques – building on the foundations of historical materialism – that we are most in need of today.
>
> (Foster and Clark, 2016, p. 22, footnote 121)

It is towards this end that third-stage ecosocialists apply Marx's method of 'ecological materialism' to analyse contemporary environmental crises and thereby advance the theoretical understanding that ecosocialist activists within the climate justice movement need in order to inform their strategies and tactics.

Third-stage ecosocialism

In her book *The Political Economy of Global Warming: The Terminal Crisis*, Del Weston (2014, p. 7) uses the analytical tools of classical Marxism that integrate metabolic rift analysis to argue that AGW 'is just one of a number of converging and accelerating symptoms of a planet plundered beyond its capacity to repair, regenerate and sustain life and civilisation as we have known it over the last 10,000 years of the Holocene epoch'. Weston (2014, p. 61) argues that 'the Marxist historical account of the development of capitalism is critical to an understanding of the political economy of global warming' for three interrelated reasons.

> First, to understand the inseparable dichotomy between accumulation and dispossession and to address this division so there can be justice for all impoverished peoples around reparations for ecological debt – a prerequisite to any solution to global warming. Second, as a basis for understanding and changing the systemic structural underpinnings of global warming. Third, to enable the development of new political economy

structures which avoid the pitfalls of capitalism. One can only understand the causes of global warming by understanding and then critiquing its political economy. Then one can begin to build alternatives that address the causes of global warming and not just those of selective symptoms.

(Weston, 2014, p. 61)

In this paragraph, Weston establishes the way in which Marxist theory links to practice, and this understanding is what informs all third-stage ecosocialist analyses of the interrelated crises engendered by the capitalist system, whose environmental destruction now extends throughout the entire Earth System (including its human component). Third-stage ecosocialist analyses thus bridge the gap between the more academic and specialist debates between first-stage and second-stage ecosocialist writings on the one hand, and activists engaged in on-the-ground political activity on the other. The distinction between theoreticians and activists is not, however, so definitive; some third-stage ecosocialist authors are not only academics but also climate justice activists, with Chris Williams being one such example. In addition to Williams's *Ecology and Socialism: Solutions to Capitalist Ecological Crisis* (2010), other popular ecosocialist texts that provide general critiques of capitalism's social and environmental harms include Joel Kovel's *The Enemy of Nature: The End of Capitalism or the End of the World?* (2007), Michael Löwy's *Ecosocialism: A Radical Alternative to Capitalist Catastrophe* (2015), Daniel Tanuro's *Green Capitalism: Why It Can't Work* (2013) and Brian Tokar's *Toward Climate Justice: Perspectives on the Climate Crisis and Social Change* (2014).[11] In addition to providing broad overviews of the way in which capitalism is inherently both exploitative and anti-ecological, another approach taken by third-stage ecosocialist authors is to focus their analysis more specifically on particular issues, as Stefano Longo, Rebecca Clausen and Brett Clark do in their book *The Tragedy of the Commodity: Oceans, Fisheries, and Aquaculture* (2015) and as Ian Angus does in *Facing the Anthropocene: Fossil Capitalism and the Crisis of the Earth System* (2016).

According to Foster (2016, p. 413), it was István Mészáros, a student of renowned Marxist Georg Lukács, who provided 'the first comprehensive Marxian critique of the emerging planetary ecological crisis in his 1971 Deutscher Prize Lecture – published a year before the Club of Rome's *Limits to Growth*'.[12] In this lecture, Mészáros 'argued that the waste-based accumulation characterizing US monopoly capitalism could not be expanded globally without breaking the ecological budget of the entire planet' (*ibid.*). While Mészáros' work is not easily accessible to a non-academic audience due to the complexity of the concepts he discusses, Ian Angus' book on the Anthropocene provides a very clear account of the implications of the planetary metabolic rift that has resulted in this new geological epoch and, most importantly, what responsibilities these developments place on ecosocialists. The exemplary way in which Ian Angus has explained the Anthropocene is evident in the fact that Clive Hamilton, who contends that most people simply do not understand this new geological epoch's radical implications, has praised

Angus not once but twice in his book *Defiant Earth: The Fate of Humans in the Anthropocene* (Hamilton, 2017) for conveying these implications so clearly.[13] Moreover, the value of Angus' work in publishing this book is also evident in the many ecosocialist discussions it has initiated. In addition to writing books that are easily accessible to the general public and are also widely read by activists, ecosocialists publish articles in various media and on a variety of topics, including on current events. They also participate in climate movement activist debates about strategies and tactics that are appropriate for supporting and furthering the project of achieving radical climate justice, as discussed in more detail in Chapter 6.

Notes

1 Foster (cited in Stache, 2016) identifies Paul Burkett's *Marx and Nature* (2014a) and his own *Marx's Ecology: Materialism and Nature* (2000) as other important early examples of second-stage ecosocialist works.

2 Foster points out that anti-critiques are, in themselves, also valuable exercises, as the process of conducting them constitutes a powerful method facilitating 'self-clarification and a degree of self-critique, together with a major dialectical advance in theoretical understanding. In this way Marxism has continually deepened and revolutionized its perspective, renewing itself in terms of both its foundational views and new historical challenges' (Foster, cited in Stache, 2016).

3 Herman Daly is an early founder of a tendency within the environmental movement that Leahy characterises as 'radical reformism' (Leahy, 2017, p. 1). While this group of theorists critiques some features of capitalism for causing environmental damage, according to Leahy (p. 2) they 'primarily identify as environmentalists, rather than as socialists or anarchists'. Unlike ecological Marxists, they do not see capitalism (as a system) as *inherently* anti-ecological. Despite the radical nature of the reforms they propose to address the issue of sustainability (including a steady-state economy, the redistribution of wealth, strong government regulation and instituting minimum and maximum income levels), they do not propose the abolition of capitalism because of their belief that 'properly functioning markets allocate resources efficiently' (Leahy, 2017, p. 5). The reforms they propose are so radical, however, that Leahy argues that they would likely lead to socialism if implemented (which is why the capitalist class and its political representatives and defenders are unlikely to countenance instituting them, as some of these theorists hope they will do voluntarily). Leahy (2017, p. 13) concludes that because of its radical potential, 'radical reformism is not just ecological modernization in a new garb'. Nevertheless, the radical reformist perspective is seen as problematic by ecological Marxists; Leahy outlines one Marxist critique of this perspective as being that it 'gives credibility to the very structures which "have been responsible for environmental decline" in the first place' (Leahy, 2017, p. 13).

4 Some of the original works that Burkett (2005) refers to in his responses to specific criticisms of Marx and Engels' ideas include *Capital*, Volumes 1, 2 and 3; *Critique of the Gotha Programme*; *Economic and Philosophical Manuscripts of 1844*; *Anti-Dühring*; *The German Ideology*; *Dialectics of Nature*; *Grundrisse*; and *Theories of Surplus Value*.

5 Refer to Burkett (2005, 2014a, 2014b) for many more detailed investigations of the works of Marx and Engels demonstrating that, contrary to critics' assertions, their vision of communism is inherently environmentally sustainable.

6 Refer to Clapp (2009) for a detailed analysis of features of the neoliberal global economy, including financial speculation in commodities, which led to sharp price hikes in basic food staples in 2008. Clapp's analysis reveals structural features within the global economy that are likely to lead to more food price hikes that further disadvantage the poorest and most vulnerable people in the world while generating profits for wealthy institutional investors.

7 Refer to Shellenberger and Nordhaus (2004) and the Breakthrough Institute document *An Ecomodernist Manifesto* (Asafu-Adjaye et al., 2015) for original accounts of the arguments this group of 'ecomodernists' present.

8 Refer to Angus (2015) for the way in which ecomodernist attempts to appropriate the meaning of the Anthropocene also aim 'to sow confusion by promoting a caricature that has nothing to do with the actual Anthropocene and everything to do with preserving the status quo'.

9 Foster (2016, p. 414) cautions that the complex method of historical materialism developed and deployed by Marx and Engels should not be confused with 'the dogmatic, mechanical views that were sometimes crudely advanced in the Soviet Union under this label'.

10 In the extracts referred to here, Foster and Clark (2016) reference Marx's *Text on Methods*, *Economic and Philosophical Manuscripts*, *Poverty of Philosophy*, and *Capital*, Volumes 1 and 3.

11 This is just a very small and randomly selected sample of the many ecosocialist monographs that have been written. Refer to the ecosocialist website climateandcapitalism.com for regular reviews of recently published ecosocialist monographs and other books relevant to the crises we face.

12 The Deutscher Prize is awarded annually to the author of 'the book which exemplifies the best and most innovative new writing in or about the Marxist tradition'; István Mészáros received the Prize in 1970 for his book *Marx's Theory of Alienation* (www.deutscherprize.org.uk/wp/).

13 Hamilton (2017) describes Angus's overview of the science as 'superb' (p. 10) and later says 'Ian Angus presents a Marxist view of the Anthropocene that, whatever one may think of the politics, stays true to the new science of the Earth System' (p. 20).

References

Angus, I. (2011). How to Make an Ecosocialist Revolution. green left, 7 October. Available at www.greenleft.org.au/content/how-make-ecosocialist-revolution-watch-video.

Angus, I. (2015). Hijacking the Anthropocene. Climate & Capitalism, 19 May. Available at http://climateandcapitalism.com/2015/05/19/hijacking-the-anthropocene.

Angus, I. (2016). *Facing the Anthropocene: Fossil Capitalism and the Crisis of the Earth System*. New York: Monthly Review Press.

Angus, I. and Foster, J.B. (2016). In Defence of Ecological Marxism: John Bellamy Foster Responds to a Critic. Climate & Capitalism, 6 June. Available at http://climateandcapitalism.com/2016/06/06/in-defense-of-ecological-marxism-john-bellamy-foster-responds-to-a-critic.

Asafu-Adjaye, J., Blomqvist, L., Brand, S., Brook, B., Defries, R., Ellis, E., Foreman, C., Keith, D., Lewis, M., Lynas, M., Nordhaus, T., Pielke, R., Jr., Pritzker, R., Roy, J., Sagoff, M., Shellenberger, M., Stone, R. and Teague, P. (2015). *An Ecomodernist Manifesto*. The Breakthrough Institute. Available at www.ecomodernism.org/manifesto-english.

Baines, J. (2017). Accumulating through Food Crisis? Farmers, Commodity Traders and the Distributional Politics of Financialization. *Review of International Political Economy*, 24(3), 497–537.

Buck, C.D. (2013). Post-Environmentalism: An Internal Critique. *Environmental Politics*, 22(6), 883–900.

Burkett, P. (2005). Marx's Vision of Sustainable Human Development. *Monthly Review*, 57(5), 34–62.

Burkett, P. (2014a). *Marx and Nature: A Red and Green Perspective*. Chicago, IL: Haymarket Books.

Burkett, P. (2014b). Two Stages of Ecosocialism? *International Journal of Political Economy*, 35(3), 23–45.

Clapp, J. (2009). Food Price Volatility and Vulnerability in the Global South: Considering the Global Economic Context. *Third World Quarterly*, 30(6), 1183–1196.

Clark, B. and Foster, J.B. (2010). Marx's Ecology in the 21st Century. *World Review of Political Economy*, 1(1), 142–156.

Fletcher, R. (2012). Capitalizing on Chaos: Climate Change and Disaster Capitalism. *ephemera: theory & politics in organization*, 12(1/2), 97–112.

Foster, J.B. (1992). The Absolute General Law of Environmental Degradation under Capitalism. *Capitalism Nature Socialism*, 3(3), 77–81.

Foster, J.B. (2000). *Marx's Ecology: Materialism and Nature*. New York: Monthly Review Press.

Foster, J.B. (2002). II. Capitalism and Ecology: The Nature of the Contradiction. *Monthly Review*, 54(4), 6–16.

Foster, J.B. (2009). *The Ecological Revolution: Making Peace with the Planet*. New York: Monthly Review Press.

Foster, J.B. (2014). Foreword. In Burkett, P., *Marx and Nature: A Red and Green Perspective*. Chicago, IL: Haymarket Press.

Foster, J.B. (2015). Marxism and Ecology: Common Fonts of a Great Transition. The Great Transition Initiative, October. Available at https://greattransition.org/publication/marxism-and-ecology.

Foster, J.B. (2016). Marxism in the Anthropocene: Dialectical Rifts on the Left. *International Critical Thought*, 6(3), 393–421.

Foster, J.B. and Clark, B. (2016). Marx's Ecology and the Left. *Monthly Review*, 68(2), 1–25.

Foster, J.B., Clark, B. and York, R. (2010). *The Ecological Rift: Capitalism's War on the Earth*. New York: Monthly Review Press.

Hamilton, C. (2013). *Earth Masters: Playing God with the Climate*. Crows Nest, NSW: Allen & Unwin.

Hamilton, C. (2015). Getting the Anthropocene So Wrong. *The Anthropocene Review*, 2(2), 102–107.

Hamilton, C. (2017). *Defiant Earth: The Fate of Humans in the Anthropocene*. Crows Nest, NSW: Allen & Unwin.

Harvey, D. (2005). *The New Imperialism*. New York: Oxford University Press.

Holleman, H. (2015). Method in Ecological Marxism: Science and the Struggle for Change. *Monthly Review*, 67(5), 1–10.

Hudis, P. (2012). *Marx's Concept of the Alternative to Capitalism*. Chicago, IL: Haymarket Books.

Klein, N. (2007). *The Shock Doctrine: The Rise of Disaster Capitalism*. London: Penguin.

Klein, N. (2017). How Power Profits from Disaster. *The Guardian*, 6 July. Available at www.theguardian.com/us-news/2017/jul/06/naomi-klein-how-power-profits-from-disaster.

Kovel, J. (2007). *The Enemy of Nature: The End of Capitalism or the End of the World?* New York: Zed Books.

Latour, B. (2015). Fifty Shades of Green. *Environmental Humanities*, 7(1), 219–225.

Leahy, T. (2017). Radical Reformism and the Marxist Critique. *Capitalism Nature Socialism*, 29(2), 61–74.

Longo, S.B., Clausen, R. and Clark, B. (2015). *The Tragedy of the Commodity: Oceans, Fisheries, and Aquaculture*. New Brunswick, NJ: Rutgers University Press.

Löwy, M. (2015). *Ecosocialism: A Radical Alternative to Capitalist Catastrophe*. Chicago, IL: Haymarket Books.

Monbiot, G. (2015). Meet the Ecomodernists: Ignorant of History and Paradoxically Old-Fashioned. *The Guardian*, 24 September. Available at www.theguardian.com/environment/georgemonbiot/2015/sep/24/meet-the-ecomodernists-ignorant-of-history-and-paradoxically-old-fashioned.

Pike, A. and Pollard, J. (2010). Economic Geographies of Financialization. *Economic Geography*, 86(1), 29–51.

Randalls, S. (2010). Weather Profits: Weather Derivatives and the Commercialization of Meteorology. *Social Studies of Science*, 40(5), 705–730.

Shellenberger, M. and Nordhaus, T. (2004). *The Death of Environmentalism: Global Warming Politics in a Post-Environmental World*. Available at www.thebreakthrough.org/images/Death_of_Environmentalism.pdf.

Stache, C. (2016). Marx and the Earth: Why We Wrote an 'Anti-Critique'. Climate & Capitalism, 22 August. Available at http://climateandcapitalism.com/2016/08/22/marx-and-the-earth-why-we-wrote-an-anti-critique.

Szerszynski, B. (2015). Getting Hitched and Unhitched with the Ecomodernists. *Environmental Humanities*, 7(1) 239–244.

Tanuro, D. (2013). *Green Capitalism: Why It Can't Work*. London: The Merlin Press Ltd.

Tokar, B. (2014). *Toward Climate Justice: Perspectives On the Climate Crisis and Social Change*. Porsgrunn, Norway: New Compass Press.

Weston, D. (2014). *The Political Economy of Global Warming: The Terminal Crisis*. New York: Routledge.

Williams, C. (2010). *Ecology and Socialism: Solutions to Capitalist Ecological Crisis*. Chicago, IL: Haymarket Books.

4 Institutional responses to a changing biosphere

The Intergovernmental Panel on Climate Change and the United Nations Framework Convention on Climate Change

Introduction

The discovery of anthropogenic global warming (AGW) elicited a number of official responses, including the establishment of a set of institutions to address the dangers it poses. In this chapter, the origins and evolution of the two primary institutions established to deal with climate change, the Intergovernmental Panel on Climate Change (IPCC) and the United Nations Framework Convention on Climate Change (UNFCCC), are discussed with reference to interactions at the 'world order' and 'forms of state' levels. The analysis in this chapter demonstrates the validity of ecosocialist claims that the institutional arrangements making up the official climate change 'regime' are incapable of achieving their stated aim of avoiding dangerous climate change (for example, refer to the ecosocialist positions represented in the writings of Angus, 2007; Kovel, 2007; Longo, Clausen and Clark, 2015; Tokar, 2014; Williams, 2010).[1]

Ecosocialists and other climate justice activists and advocates focus most of their critique of the formal climate change regime mechanisms on the serious inadequacies of the UNFCCC's outcomes, particularly since the fifteenth Conference of the Parties (COP-15) convened in Copenhagen in 2009. Like other analysts writing about official responses to climate change, ecosocialists widely (and appropriately) cite the content of IPCC reports on the physical science of climate change to corroborate their evaluations of the severity of AGW and the urgent need to take immediate effective action in order to mitigate further warming. Ecosocialist discussions of the IPCC include critiques of this institution, particularly related to its inherent tendencies to err on the side of conservatism (Angus, 2007). This chapter builds on ecosocialist critiques of this intergovernmental scientific body by discussing its origins and evolution and demonstrating that the entire climate change regime (including the IPCC) was designed to forestall and prevent socially just and ecologically benign solutions to anthropogenic global warming. Far from seeking to criticise the many scientists who volunteer their time and services (often at great personal cost) to produce the IPCC assessment reports, the aim of the analysis presented in this chapter is rather to demonstrate that these scientists work within a context that is designed to constrain the use of scientific evidence to support rational policymaking in achieving anthropogenic GHG emission reductions and reorganising social relations of production appropriately.[2] By way of introducing the key issues and actors involved, the discussion begins with how the first US Bush Administration responded to a prominent scientist's testimony about the dangers posed by AGW.

Official responses: constructing social facts about anthropogenic global warming

On 23 June 1988, Dr James Hansen, then-head of the US National Aeronautics and Space Administration (NASA) Goddard Institute for Space Studies, made a landmark testimony before the United States Senate Committee on Energy and Natural Resources. In this testimony, Hansen informed the Reagan Administration that the accumulation of anthropogenic GHGs in the Earth's atmosphere was causing dangerous global warming due to the greenhouse effect, and that this was leading to climate change (Baer, 2014). It is well documented, however, that scientific investigations into the relationship between GHGs and the Earth's average temperature date back to the nineteenth century. In 1824, Joseph Fourier first proposed that the Earth's atmosphere traps some of the heat radiating back into space from the Earth's surface and in 1861, John Tyndall identified the gases responsible for this effect; in 1896, Svante Arrhenius' calculations estimated that a doubling of CO_2 in the atmosphere from burning fossil fuels would raise average global temperatures by 5 °C to 6 °C (which is similar to the range estimated by today's climate scientists), although he did not foresee the exponential rate at which CO_2 concentrations would increase – a fact demonstrated by the now-famous 'Keeling Curve' that plots the continuous measurements of CO_2 levels in the atmosphere, which began under the supervision of Charles David Keeling in the 1950s (Bodansky, 2001; Union of Concerned Scientists, 2012; World Bank, 2014). In 1965, when US President Johnson asked his President's Science Advisory Committee to report on the potential problems of environmental pollution, the Committee's report included a 23-page appendix providing the first official warning that CO_2 emissions from the burning of fossil fuels 'could rapidly reshape Earth's climate' (Peterson, Connolley, and Fleck, 2008; see also Agrawala, 1998a). One major uncertainty debated amongst climate scientists after 1965 was the extent to which aerosol cooling from pollutant particles in the atmosphere was counteracting greenhouse warming, but by 1978 James Hansen and his team of researchers had concluded that 'greenhouse warming had become the dominant forcing' (Peterson, Connolley and Fleck, 2008, p. 1329), a result confirmed by numerous subsequent scientific investigations.

While Hansen was demonstrably not the first scientist to link global warming to the accumulation of anthropogenic GHG emissions in the Earth's atmosphere, this testimony has been described as a pivotal point in contemporary environmental politics in that it brought the issues of global warming and climate change to the attention of policymakers, the media and the public (Hecht and Tirpak, 1995; Milman, 2015). A conjuncture of extreme weather events in 1988, including it being the hottest US summer on record, made Hansen's testimony particularly salient (Armitage, 2005; Agrawala and Andresen, 1999; Bodansky, 2001). As Armitage recounts:

> [t]he world seemed to get a glimpse of its climate future in 1988. Heat and drought caused severe crop losses in the American Midwest, the worst since the Dust Bowl of the 1930s. The USSR suffered drought, as did China. Unexpected floods ravaged Africa, Brazil, Bangladesh and India. Hurricanes struck the Caribbean, along with a cyclone in New Zealand and a typhoon in the Philippines. These disasters provided compelling background to the political events of 1988, the most dramatic year in the history of the politics of climate change.
>
> (Armitage, 2005, p. 420)

The incoming administration of George H.W. Bush (1989–1993) responded by attempting to prevent Hansen and other US climate scientists from communicating their results, restricting media access to climate scientists and editing major reports on climate change and systematically ensuring that information from such reports was not used in other government policy documents (Armitage, 2005; Rich and Merrick, 2007).

Legal scholars Rich and Merrick (2007, pp. 243–244) point out that the Bush Administration's control of information about climate science did not violate US federal laws and regulations, and that it used three key legal frameworks to more directly control the dissemination of scientific information to the public: implementation of the Data Quality Control Act in 2001; the establishment of a centralised peer-review process directly under White House control in 2004; and the granting of greater powers to federal agencies 'to designate material as "classified" and "sensitive but unclassified"' (Rich and Merrick, 2007, pp. 243–244). The legality of the Bush Administration's measures to restrict the dissemination of information about climate change demonstrates the way in which liberal democratic institutions provide tools that can be (and frequently are) used to prevent addressing serious issues that threaten vested interests. The Bush Administration's use of legislation to restrict the dissemination of information about climate change also demonstrates the validity of ecosocialist arguments that addressing the threat posed by anthropogenic global warming calls for the radical reorganisation of society. This argument is further strengthened when one considers that the US has not been alone in responding to scientific findings that threaten business interests by restricting funding for climate science on the one hand and the public's right to information about the findings of publicly funded science on the other.

Like the Bush Administration, the conservative Harper government (2006–2015) in Canada also 'gutted environmental legislation, terminated environmental monitoring programs, muzzled government scientists, and laid off over

2000 researchers from federal labs' (Pelley, 2015, p. 528). While the current Trudeau Liberal Party government has reversed some of these measures, other policies (such as the Harper government's 20% reduction of government science department budgets) present a longer-term erosion of the Canadian climate science research effort (*ibid.*). In Australia, the conservative Coalition Howard government (1996–2007) refused to sign the Kyoto Protocol, and the Australian Labor Party Rudd/Gillard government (2007–2013) measures supporting climate science and renewable energy research and pursuing GHG reduction targets by implementing a carbon pricing mechanism were aggressively reversed by the conservative Coalition government (2013–present). Under the leadership of Tony Abbott, the Coalition government (2013–2015) dismantled and defunded organisations conducting climate change research and repealed the carbon pricing legislation implemented by the Labor government in 2012 (Beeson and McDonald, 2013; Crowley, 2017). The Turnbull Coalition government (2015–2018) continued to defund climate research and to both defund renewable energy projects and use regulations to block their development (Bainbridge, 2017; Swann, 2016); it simultaneously supported large-scale fossil fuel projects through its 'aggressive promotion of coal' (Morgan, 2017) and policies that facilitated the expansion of natural gas and coal-seam gas industries (Baer, 2016). Coalition government policies promoting the extraction and use of fossil fuels in Australia intensified after the 'leadership spill' in the Liberal Party that replaced Malcolm Turnbull with Scott Morrison (2018–present). Government regulations obstructing effective action to mitigate climate change and promoting fossil fuel industries are complemented by a variety of other tactics that some officials, policymakers, media outlets and other representatives and supporters of powerful vested interests have long resorted to. These tactics include denying the reality of AGW, downplaying the seriousness of the effects of GHG emissions, or even claiming that these effects are beneficial, and casting doubt on whether global warming is anthropogenic – claiming, instead, that it is part of a 'natural cycle' (Oreskes and Conway, 2010; Peterson, Connolley and Fleck, 2008).

Despite these misrepresentations of the evidence, there has been a growing scientific consensus since at least the 1980s that anthropogenic global warming is real and that its dangerous effects are already playing out. There is also a widespread scientific consensus that it is necessary to decarbonise the global economy as soon as possible (Åhman, Nilsson and Johansson 2016) while planning adaptation strategies for dealing with the warming and the effects of climate change that are now inevitable. These concerns about global warming and climate change initially arose within the context of growing awareness and concerns about a wider environmental crisis that had prompted the convening of the 1972 United Nations Conference on the Human Environment in Stockholm, the first of many 'megasummits' emphasising the need for 'sustainable development' (Biermann, 2013). Ineffective international institutional responses to this wider environmental degradation are summarised below and again confirm ecosocialist arguments that capitalism and its political institutions cannot be reformed to deal with the environmental and climate change crises that they engender in their normal operations.

International institutional responses to environmental degradation: 1972–present

According to several analysts, the publication of Rachel Carson's *Silent Spring* in 1962 found a receptive audience amongst the public both in the US and worldwide and heralded the beginning of the modern environmental movement in the Global North (Sale, 1993).[3] Widespread public concern about environmental degradation was prompted by a variety of issues, including fears of nuclear fallout from the atmospheric testing of nuclear bombs, a growing awareness of the dangers to human health and security posed by the pollution caused by the post-war economic boom, several highly publicised environmental disasters and early predictions of an approaching 'doomsday' scenario as a result of ecological collapse if industrialisation continued on a trajectory of infinite growth on a finite planet (Sale, 1993). Sale (1993, pp. 39–40) identifies the formation of the Club of Rome in 1968 as 'the first significant international recognition of the environmental crisis'.

The Club of Rome commissioned a Massachusetts Institute of Technology (MIT) team to conduct a research project that used a complex computer model to analyse global economic and environmental trends. Publishing their findings 'to considerable fanfare' in the report *Limits to Growth* in March 1972, the MIT researchers predicted catastrophe sometime during the twenty-first century.

> If the present growth trends in world population, industrialization, pollution, food production, and resource depletion continue unchanged, the limits to growth on this planet will be reached sometime within the next 100 years. The most probable result will be a rather sudden and uncontrolled decline in both population and industrial capacity.
>
> (Meadows, Randers and Meadows, 2004, front pages)

While the Club of Rome was a private coalition of prominent scientists, politicians, technocrats and businessmen from 25 countries, the first official international response to environment concerns by the UN was the Biosphere Conference in Paris in 1968, where it was agreed to hold a UN Conference on the Human Environment (UNCHE) in 1972 (Sale, 1993). One important outcome of this meeting, which came to be known as the 'Stockholm Conference', was the decision to establish the United Nations Environment Program (UNEP), albeit as a body coordinating its work through other agencies (rather than as a specialised agency). The UNEP's 'paltry budget' and subordinate role within the UN system meant that it had no enforcement powers with which to act on its mandate of coordinating 'all matters on global ecosystems' (*ibid.*, pp. 42–43). In addition to establishing the UNEP, another UNCHE outcome was the adoption of the Stockholm Declaration, containing 26 principles concerning the environment and development (UNGA, 1972). Despite this declaration, and the Stockholm Action Plan with its 109 recommendations, the global spread of capitalist relations of production (generally referred to as 'development' in official documents and popular accounts) continued to damage the environment and in 1983 the Secretary General of the UN requested Gro Brundtland, the former Prime Minister of Norway, to establish and chair an independent commission on environmentally sustainable development (WCED, 1987). The official title of this organisation was the World Commission on Environment and Development (WCED), but it is also widely referred to as the 'Brundtland Commission'. *Our Common Future* (1987), the outcome of the Brundtland Commission, is a key document often cited in the sustainability literature for its emphasis on inter-generational equity (the need to protect the environment that future generations will inherit) and the view that 'sustainable development' can be achieved through a balance between economic growth, environmental protection and social equity – the three 'key pillars' of sustainable development (Brundtland, 1987; WCED, 1987). These initial attempts to mobilise the 'global community' of the world's governments to take effective action on addressing the many issues of environmental degradation posed by the global expansion of the capitalist mode of production were followed by numerous other initiatives and conferences, including the 1992 United Nations Conference on Environment and Development (UNCED) convened in Rio de Janeiro, also known as the first 'Earth Summit'. The first Earth Summit was followed by the 2002 'Rio+10' Summit and then by the 2012 United Nations Conference on Sustainable Development (UNCSD) (Biermann, 2013).[4]

Since the Brundtland Commission's findings, climate change has also featured explicitly as an issue to be addressed at these environmental summits (Boehmer-Christiansen, 1994), but while there have been many 'statements of principles' and 'action plans', the goal of 'sustainable development' remains elusive (Hadden and Seybert, 2016). The global spread and intensification of capitalist relations of production continues to degrade the environment and compromise the integrity of the biosphere at an ever-increasing rate nearly five decades after the emergence of these issues. US President George H.W. Bush's statement that 'the American way of life is not negotiable' (cited in Harris, 2009, p. 968) at the first Earth Summit clarified that US policy on international environmental protection measures took a backseat to US economic interests, and successive US administrations (whether Republican or Democrat) continue to protect the short-term interests of US capital (Falkner, 2005).

The 2012 Earth Summit outcome was particularly disappointing to the scientists who had participated in the preparatory 2012 Planet Under Pressure: New Knowledge Towards Solutions conference held in London to prepare for the Rio+20 Summit (Biermann, 2013). The scientists' urgent statements about the severity of the environmental emergency are published in the State of the Planet Declaration (DIVERSITAS and ICSU, 2012). Due to resistance from a number of countries, but particularly from the United States (Biermann, 2013), like all previous scientists' warnings these, too, failed to initiate the structural changes required within the UN system to strengthen the 'environmental pillar' of sustainability, or to produce a binding agreement that would lead to effective action that may avert planetary disaster as a result of global warming. Siding with some developing nations (a tactic often resorted to by US government representatives), the United States rejected the 'notion of planetary boundaries' at the Rio+20 Conference, and thus also the related reform proposal that NGOs put forward to establish a 'United Nations High Commissioner for Future Generations' (Biermann, 2013). Other reforms proposed to strengthen the UN's ability to protect the environment, such as the creation of a 'World Environmental Organisation' as a Specialised Agency to eliminate the inconsistencies between different environmental programmes and organisations working on specific issues often at odds with one another, and the integration of UN environmental, economic and social policies under a body such as a 'Sustainable Development Council' were also resisted by some governments, and were strongly resisted by US representatives (Biermann, 2013).[5] This response is no different to the ineffectiveness of official responses to the climate change crisis; the restricted mandates, limited autonomy and lack of enforcement powers of the formal international and intergovernmental institutions nominally established to address the issues of GHG emissions, AGW and climate change ensure that 'business as usual' — particularly a world economy based on the burning of fossil fuels — continues irrespective of the environmental and social costs of this trajectory.

Many analysts identify the United States as being at the forefront of efforts by various governments and groups representing vested interests to ensure that global institutions have no legislative or executive authority and power to take the measures required to mitigate climate change. However, some of the ways in which the US has shaped the international institutions dealing with climate change are not widely discussed in contemporary accounts of climate politics. This history is significant and can be reclaimed if one refers to work published in the 1990s, soon after the institutions tasked with addressing AGW had been established. Shardul Agrawala (1998a, 1998b, 1999) provides detailed accounts of the origins and early evolution of two of these institutions, the Advisory Group on Greenhouse Gases (AGGG) and the IPCC, and these works are referred to extensively in relevant sections of this chapter.

World order, material capabilities, institutions and competing ideas: the origins, establishment and evolution of international and intergovernmental institutions to address global warming

Concerns provoked by a growing awareness of the potential danger of accumulating GHGs in the atmosphere are reflected in numerous collaborations between scientists that began as early as 1967 with the establishment of the Global Atmospheric Research Program under the auspices of the World Meteorological Organization (WMO) and the International Council for Science (ICSU) (Agrawala, 1998a; ICSU, 2015). These collaborations were initially mobilised through loose research networks and a number of conferences in the 1970s before the WMO, the United Nations Environment Program (UNEP) and the ICSU organised and convened the first World Climate Conference (WCC) in 1979 (Agrawala, 1998a). The WCC held a series of international atmospheric science workshops in Villach (Austria) in the 1980s (Agrawala, 1998a; Gupta, 2010). In the third workshop (1985), widely known as the Villach Conference, the majority consensus amongst the participating scientists was that by the first half of the twenty-first century, the Earth's mean global temperature would be greater than at any time in human history, and they recommended further research in order to clarify the nature of the threat (Zillman, 2009). The Villach Conference findings were summarised in a report entitled *The Greenhouse Effect, Climatic Change and Ecosystems* (1986), which the ICSU (2015, p. 15) describes as 'the first comprehensive international assessment of the environmental impact of atmospheric greenhouse gases'. This report, published more than three decades ago, concluded that the accumulation of CO_2 in the atmosphere is caused by 'human activities' and warned that a doubling of CO_2 would cause 'substantial warming', recommending international cooperation on 'a variety of specific policy actions' to address the issue (ICSU, 2015). The Villach Conference also led to the WMO, the UNEP and the ICSU establishing the first body of international experts to guide climate policy, the AGGG. The AGGG's history and swift demise are discussed in some detail below as this case study illustrates the US government's activism, facilitated by its vast material capabilities and its leading position in maintaining the current neoliberalising capitalist world order, in blocking any measures to effectively address global warming from the very beginning of the development of a widespread awareness of this issue.[6]

The Advisory Group on Greenhouse Gases

It is significant that it was UNEP Director Mostafa Tolba, who had played a central role in the development and adoption of the Vienna Convention for the Protection of the Ozone Layer, who first introduced the idea of establishing an advisory panel to guide climate policy (Agrawala, 1999). Following Tolba's suggestion in his opening address at the 1985 Villach Conference, the WMO, UNEP and ICSU established the AGGG for this purpose in July 1986 (Agrawala, 1999). Participants in a workshop held in Bellagio in November 1987 proposed that the AGGG 'design policies aimed at limiting increases in temperature and sea level to within "tolerable rates"' – a proposal that constitutes what Agrawala (1999, p. 162) identifies as 'the first explicit *policy* debate on climate change' (emphasis in original). The AGGG 'Toronto

Conference on the Changing Atmosphere' that was held in June 1988 resulted in the Toronto Declaration, which was unprecedented in that it called 'for a 20% reduction in OECD GHG emissions from 1988 levels by 2005 [which] made it the most significant policy initiative of its time on climate change' (Agrawala, 1999, p. 163). The rapid succession of policy messages on action to address climate change that emerged after the Bellagio workshop, the Brundtland Report and the Toronto Conference drew international attention to the issue of climate change, and it was at about this time (in 1988) that the US government 'began to flex its muscle on the international arena, although efforts in that direction had begun at least two years earlier' (Agrawala, 1999, p. 164).

Agrawala (1999) notes the complex and contradictory role of the US in the context of growing awareness of the climate change issue; while US scientists and scientific institutions and agencies produced most of the scientific knowledge about AGW, the political and economic interests threatened by taking the action necessary to stop the anthropogenic GHG emissions causing it were 'huge' for the US. At that time the US was the largest GHG emitter, and the powerful fossil fuel and car manufacturing lobbies had the active support of a Republican White House to prevent the adoption of binding agreements limiting fossil fuel use (Agrawala, 1999).[7] In addition to government and industry resistance to taking actions that could 'damage' the US economy, government agencies had different views on the severity of global warming and on whether or not the issue warranted a policy response (Hecht and Tirpak, 1995).

Disagreements between different government agencies about the severity of AGW were already evident in 1986, when the US National Climate Program Policy Board was convened to discuss a letter addressed to Secretary of State George Schultz by UNEP Director Mostafa Tolba (Agrawala, 1999). Tolba had requested help in initiating the process for a climate convention, and the government agency representatives participating in these discussions agreed to the US recommendation for the establishment of an ' "*intergovernmental mechanism*" … to conduct scientific assessment of climate change' (Agrawala, 1999; emphasis in original). Agrawala (1999, p. 164) notes that all parties at this meeting shared 'a common concern with respect to the international policy initiative that had begun with the 1985 [AGGG] Villach workshop'. While those who thought that climate change was a serious issue were opposed to 'a small set of "free wheeling experts" ' (in other words, scientists) formulating international policy, advocates of delaying the adoption of any measures to restrict GHG emissions were concerned that pressure from expert groups like the AGGG might force the US government into 'premature policy commitments' (Agrawala, 1999, pp. 164–165). Agrawala (1998a) thus makes the important point that analysts who present *sequential* accounts of the IPCC's establishment as a result of the findings and discussions at the 1987 AGGG workshops and the 1988 Toronto Conference are mistaken, and that the IPCC was established as an *alternative* to the AGGG and to its prior activities and recommendations.

[T]he process to set up the IPCC was in motion as early as 1986, and the WMO Executive Council resolution to this effect was passed in June 1987, a few months *before* the Villach/Bellagio workshops, and a full year *before* the Toronto Conference and the hot summer of 1988. These events clearly had no role in the decision to set up the IPCC ... Instead, the trigger for the IPCC was the activism by Mostafa Tolba, the dissatisfaction in the US about the AGGG, and sharply different views on climate change amongst various US government agencies and the White House administration. The subsequent shape the IPCC took reflected a common denominator agreement between various US agencies. Reportedly there were also strategic attempts both by WMO and the US to prevent Mostafa Tolba from 'capturing' climate, the way he had, ozone [sic].

(Agrawala, 1998a, p. 612; emphasis in original)

US representatives participating in the Montreal Protocol on Substances that Deplete the Ozone Layer negotiations had provoked the ire of the US Departments of Energy, Commerce and the Interior, the Council of Economic Advisers and the Office of Management and Budget, with advisers in these departments claiming that the Environmental Protection Authority (EPA) and the State Department had acted 'too aggressively' in the ozone treaty negotiations, and without consulting other agencies (Agrawala and Andresen, 1999, pp. 471–472).[8] To curb the power of the EPA and the State Department, the White House Domestic Policy Council was given direct control over all international environmental negotiations in 1987 (*ibid.*). White House advisers representing economic and business interests thus 'out-muscled' the EPA and the State Department when it came to the George H.W. Bush Administration deciding what to do about global warming in 1989, with the Council of Economic Advisers projecting 'significant costs of various policy measures to mitigate climate change' and other senior advisers expressing concerns 'about potential US financial obligations toward developing countries under a climate convention' (Agrawala and Andresen, 1999, p. 472). While these domestic pressures explain the first Bush Administration's decision to establish an *intergovernmental* mechanism rather than a group of scientific experts to investigate and address the issue of climate change, this does not explain why other actors acquiesced in this proposal.

Agrawala (1998a) identifies several reasons that led to WMO and UNEP member governments supporting the US in establishing the IPCC: 'US clout' in the decision-making of international institutions such as the WMO and UNEP; the widespread economic implications of the policies needed to address climate change (see also Boehmer-Christiansen, 1994); and the fact that the issue of climate change was already politicised.

[T]he international environmental arena was already politicized *because climate change came in the wake of ozone.* Peter Usher, Tolba's key advisor during the ozone negotiations, admits that the ad-hoc, low key, science-driven (if politically undemocratic) nature of the early ozone assessments which led to the Vienna Convention could not be duplicated in climate change. This is because while 'politics caught up with ozone, climate change was born in politics'.

(Agrawala, 1998a, p. 614; emphasis in original)

To summarise, the IPCC acquired its present form as the result of what Agrawala describes as 'a back-room effort of design, negotiation and compromise' between US agencies in a process much of which 'is still shrouded in mystery'; it 'was the product of an intensely *political* process within the US and the UN system' (Agrawala, 1998a, pp. 615, 617; emphasis in original). US determination to play a leading role in shaping the IPCC is also evident in the fact that the US government sent 24 delegates to its first meeting in November 1988, greatly outnumbering the number of delegates from other countries – for example, Germany only sent two delegates (Boehmer-Christiansen, 1994). The US government and its allies thus ensured that the IPCC was established in a way that suits powerful corporate interests.

The Intergovernmental Panel on Climate Change

The IPCC's evolution after 1988 increasingly marginalised the roles of the WMO and UNEP while US influence increased considerably through its scientists and bureaucrats as well as through new actors such as the Global Climate Coalition and the Climate Council, which represented US fossil fuel interests and also aligned with the interests of other oil-producing economies (Agrawala, 1998b). The IPCC's role in providing scientific data informing international climate change treaties led to attempts by these actors to discredit the IPCC and its reports in the lead-up to the 1992 Rio Conference, where the UNFCCC was due to be adopted (*ibid.*). These events resulted in the IPCC overhauling its processes in an attempt to protect its legitimacy, with the first round of changes giving government representatives and international institutions such as the World Bank and the OECD a greater role in selecting report contributors and reviewers (Agrawala, 1998b). The 1993 changes also opened up the review of IPCC assessments to participating countries' 'national experts and other interested parties' and stipulated that government representatives had to approve the newly introduced 'Summary for Policymakers' (SPM) by agreeing on its contents line by line (Agrawala, 1998b; IAC, 2010). Contrasting this approach of trying 'to buy global credibility *amongst governments*' with 'the distinctly activist stance' taken by some of the IPCC's predecessors in order to 'effect prompt policy outcomes', Agrawala (1998b, p. 629; emphasis in original) reaches the surprising conclusion that 'Neither approach is implicitly superior' as both catalyzed policymakers. With the benefit of hindsight, however, it is clear that the IPCC approach has not been effective in 'catalyzing policymakers'. It is therefore reasonable to suggest that stronger policy advocacy by scientists, if it could have been achieved, may have at least partially helped mitigate some of the additional anthropogenic global warming locked in by the increasing GHG emissions in the three decades since scientists became aware of the severity of this issue.

An established intergovernmental mechanism that has now been operational for nearly two decades, the IPCC describes its main task as being to provide 'assessment reports on the state of knowledge on climate change' at regular intervals, and its restricted mandate is reflected in the statement that one of its most important principles is to produce reports that are 'policy relevant' but not 'policy prescriptive' (IPCC, 2010).[9] Rather than conducting new research, IPCC reports assess the most recently published and peer-reviewed scientific literature on climate change and related issues, but in the absence of such literature (which often is the case on issues such as adaptation), they also include information obtained from 'grey literature', which refers to government reports and work published by international organisations (IPCC, 2010). The volunteer scientists and experts conducting the assessments and writing the reports are organised into three Working Groups (WGs): WGI assesses and reports on the physical science basis of climate change; WGII focuses on the expected impacts of global warming on socio-economic and natural systems; and WGIII reports on possible policy responses to the effects identified by WGII (IAC, 2010, p. 6; Luton, 2015). This separation between the natural and social sciences is one way in which the IPCC can be controlled by governments and various other stakeholders; the physical science that provides proof that global warming is accelerating and that action to mitigate it is urgently required is the domain of WGI, while the social science discipline of economics dominates both the possible 'socio-economic' impacts and the 'policy relevant' information presented in WGII and WGIII assessment reports (Corbera et al., 2015; Hulme and Mahony, 2010). In their discussion of the practical and political decisions involved when the IPCC was being established, Hecht and Tirpak (1995, p. 385) identify WGIII's work as 'the most contentious' as it controls which 'policy relevant' information is selected for inclusion. They also draw attention to how competing views within and between US scientific and other state agencies about which WG the US should chair in the first IPCC meeting were resolved in favour of its chairing WGIII.

Aligned with this division of labour between the natural and the social sciences, the IPCC's assessment results are published in three WG reports that include 'chapters on specific topics; a Technical Summary of the chapter contents; and a Summary for Policymakers [SPM], which highlights the key findings of the assessment' (IAC, 2010, p. 8).[10] The SPMs are subject to line-by-line approval by government representatives.[11] This is a crucial mechanism that government representatives use to control the dissemination of information to the media (and, through them, to the public) as these summaries are more likely to be read given the length and complexity of IPCC reports (Hajer, 2012). But government control of the IPCC's work is far more

extensive even than this. As N.H. Ravindranath, who participated in producing eight IPCC reports, explains:

> The most powerful body of the IPCC that is responsible for making all
> the crucial decisions, starting from the contents and procedures to the
> final approval of the reports, is the 'IPCC panel' that consists of repre-
> sentatives of all the governments under the UN ... So [the] IPCC is not
> an organization with its own agenda to promote or make its own rules, it
> is continuously controlled and supervised by this panel.
>
> (Ravindranath, 2010, p. 27)

Despite its limited mandate and all the other restrictions placed on IPCC
authors, the evidence that anthropogenic GHG emissions cause global warming,
climate change and ocean acidification cannot be denied and is published in the
WGI reports. The tactic of designing the IPCC so that it has no role in recommending policies has nevertheless succeeded in delaying the adoption of the
effective policies required to mitigate global warming. Given the concern raised
within civil society by the irrefutable scientific evidence of AGW that is published in IPCC reports, however, US fossil fuel interests and their domestic and
international allies have had to resort to other tactics to block action on climate
change. These tactics include sustained and sometimes even psychologically
damaging attempts to discredit IPCC science and scientists.

Constructing competing ideas: climate change denialist attacks on IPCC climate science and climate scientists

The tactic of discrediting IPCC reports was already evident with the George
H.W. Bush Administration's rejection of the First Assessment Report in 1990,
as well as in the fossil fuel-funded Global Climate Coalition's (GCC) extensive
lobbying against climate change legislation and its 'large-scale advertising blitz
[which was] meant to assuage any trepidation the [US] public might have had
about the climate change issue' in the late 1990s (Armitage, 2005, p. 422).
There are many instances of manufactured 'climate change skepticism', or, as
the authors of the open letter 'Deniers are not Skeptics' (CSI, 2015) more
accurately describe it, 'climate change denialism'; however, three 'controversies' reported on extensively in the media and debated on social media are particularly noteworthy. These manufactured controversies have succeeded both
in damaging the reputation of IPCC science and scientists and in delaying the
action required to mitigate additional global warming (Leuschner, 2016;
McAdam, 2017) in an attempt to try to secure what Rockström et al. (2009)
refer to as a 'safe operating space' for humanity. Since one of the central functions of IPCC reports is to inform UNFCCC negotiations (Adler and Hadorn,
2014), it is not surprising that these three controversies occurred just before
important UNFCCC milestone events.

The 'hockey stick' graph controversy in the context of the Kyoto Protocol

The first controversy occurred in the lead-up to the ratification of the Kyoto Protocol and involved the 'hockey stick' graph published in the 2001 Third Assessment Report (TAR). Climate scientist Jerry Mahlman coined the term 'hockey stick graph' (Hamblyn, 2009) to describe the shape of a curve depicting changes in the Earth's temperature over several centuries (see Figure 4.1; the 'hockey stick' shape is evident on the right-hand side of the curve). Since there are no recorded temperatures for a thousand years ago, climate scientists Michael Mann, Raymond Bradley and Malcolm Hughes used a standard scientific procedure – temperature proxies obtained primarily from tree ring data – to estimate the missing data in the Northern Hemisphere temperature records and published their findings in *Nature* in 1998 (this work is commonly referred to as MBH98) and in *Geophysical Research Letters* in 1999 (MBH99) (Connolly and Connolly, 2014).

After its publication in the Third Assessment Report's WGI Summary for Policymakers, the MBH99 graph was widely publicised 'in both scientific reports and popular public presentations, and generated considerable scientific and public concern over atmospheric CO_2 concentrations' (Connolly and Connolly, 2014, p. 2). The graph's clear representation of the late twentieth

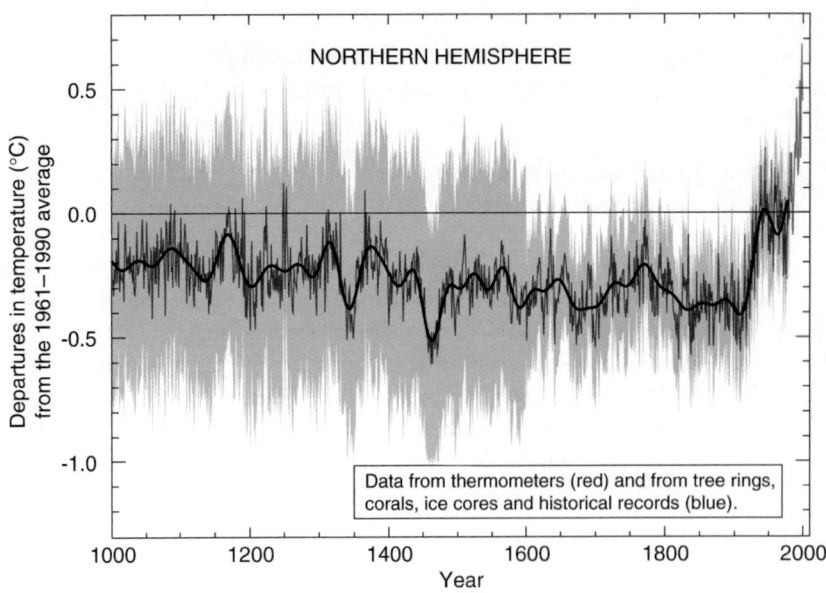

Figure 4.1 Millennial Northern Hemisphere (NH) temperature reconstruction and instrumental data from AD1000 to 1999, adapted from Mann et al. (1999).

Source: Folland et al. (2001, p. 134). © Intergovernmental Panel on Climate Change 2001. Reproduced with permission of the Licensor through PLSClear.

century's unprecedented increasing rate of global warming made it (and the scientists who developed it) the target of sustained attacks by climate change deniers (Biddle and Leuschner, 2015; Hicks, 2017). Prominent critics of the graph included Stephen McIntyre (who spent his career in the mining industry and is not a climate scientist), Ross McKitrick (an economist) and conservative US politicians such as Senator James Inhofe and Congressman Joe Barton (Biddle and Leuschner, 2015; Hicks, 2017). Although some technical flaws have been found in the 'hockey stick study', many subsequent investigations by individual researchers as well as by national and international scientific institutions confirm the 'hockey stick' graph's main message: the overall trend of a rapid increase in the rate of global warming in the latter half of the twentieth century. These investigations also clear the scientists involved of any wrongdoing (Hicks, 2017).[12] As Hicks (2017) points out, technical disagreements such as this one about the 'science' of global warming are proxies that conceal much deeper political and economic ideological disputes. The same can be said regarding the so-called 'climategate scandal' and the media publicity surrounding errors in the IPCC's Fourth Assessment Report (AR4).

COP-15, 'climategate' and AR4 errors

The timing of the leak of emails hacked from the University of East Anglia's Climate Research Unit, which occurred a few weeks prior to COP-15 in Copenhagen, makes it reasonable to suspect that this leak may have been politically motivated (Skrydstrup, 2013). A political agenda having potentially informed the leak is also perhaps evident in how, on BBC News, the lead climate negotiator of Saudi Arabia stated an expectation that this incident would 'derail the objective of the [COP-15] summit to reach a binding agreement on greenhouse gas emissions' (Skrydstrup, 2013). Widely publicised allegations that the contents of some of the leaked private emails between the scientists revealed that data had been manipulated to overstate the case of global warming and that dissenting scientific views on anthropogenic global warming had been suppressed were refuted by several official investigations in the period that followed, vindicating both the scientists and the science (Leuschner, 2016). The damage to climate science in public perceptions had, however, already been achieved with these allegations, and it was compounded with the revelation of some AR4 errors that followed 'on the heels' of this controversy (IAC, 2010, p. 2).

One of the AR4 (WGII) errors was traced back to information from a 'grey literature' source, a World Wildlife Fund (WWF) report that erroneously claimed that Himalayan glaciers were likely to disappear by 2035 (IAC, 2010). What is most controversial about this error is that the statement was questioned in peer reviewers' comments but review editors failed to address the issue (IAC, 2010). As the InterAgency Council (IAC) charged with reviewing the IPCC subsequent to these controversies points out, 'This oversight is perhaps understandable given the fact that there were a total of 37,078 review comments to deal with in the two rounds of peer review' (IAC, 2010, p. 20). Another perspective on the question of the AR4 errors is suggested by Hajer, who argues that:

> The very notion of 'errors' was, of course, not innocent in itself. That there would be some mistakes in a three volume, 3000-page-long, assessment report is not in itself necessarily a reason to problematize the quality of the science. But there was also the question of what the IPCC assessment was for. Implicit in the media outrage was the accusation that climate scientists were 'stealth issue advocates' ... scientists working for a political cause but using science as a 'fig leaf'.
>
> (Hajer, 2012, p. 460)

To restore the integrity of the IPCC's work, the IAC was charged with conducting an independent international review of IPCC policies and procedures, making recommendations that further entrench external controls on how it functions. Importantly, these recommendations include 'guidelines on who can speak on behalf of the IPCC and ... how the organization can most appropriately be represented' (IAC, 2010, p. 62), which constitutes another mechanism for controlling the dissemination of scientific information to the public. The attacks on the IPCC's alleged 'left-wing activism' from the Right of the political spectrum are compounded by politically liberal critiques arguing that its work is compromised in various ways.

Competing ideas: liberal critiques of IPCC processes and policies

Early liberal critiques of IPCC processes and outputs included: the predominance of the physical sciences (Luton, 2015) and of scientists and other experts from the Global North (Hulme and Mahoney, 2010); 'an unnecessary amplification of essentially minority opinions' used by US industry lobby groups to influence IPCC report content (Agrawala, 1998b); and the way in which acting on review comments is left to the discretion of the writing teams, who ignore 'some or a majority of all critical review comments' (*ibid.*). A particularly unjust and offensive example of how the writing team's discretion to ignore comments enabled WGIII of the Second Assessment Report to assign a cash value of $1.5 million to a human life in the OECD 'against a mere $150,000 in the developing countries' when calculating the 'social costs' of climate change (Meyer and Cooper, cited in Agrawala, 1998b, p. 626). This particular example illustrates the importance of a critique related to the composition of WGII and WGIII: the subsidiary role accorded to knowledge from social science discipline areas other than economics in these working groups (Corbera et al., 2015; Hulme and Mahoney, 2010; Luton, 2015).

The over-representation of mainstream economics theory in WGII and WGIII reports is evident in the fact that economists comprised half of the social scientists in AR5's WGII and nearly two-thirds of the coordinating lead authors of AR5's WGIII (Obermeister, 2017). This bias in the composition of WGII and WGIII is not accidental; between 1990 and 1992, US government representatives and their allies (including academics) ensured that mainstream economics would be part of the IPCC research agenda (*ibid.*). This is problematic for a variety of reasons, including that the positivist mainstream economics framings of climate change adopt a global view that excludes traditional indigenous and local knowledges (Obermeister, 2017) and results in the 'virtual invisibility' of the main victims of climate change – and especially of the poorest children (Fløttum, Gasper and St Clair, 2016).

Critics also point out that some of the sources of bias in the operations of the IPCC are not immediately evident; while the IPCC succeeded somewhat in including the participation of scientists from the Global South in AR5's WGIII, most of these scientists were trained in Global North institutions (particularly in the USA and UK), with participants from some Global South countries (such as India and Brazil) being more integrated into the IPCC research network than others (Corbera et al., 2015). Additional liberal critiques of the IPCC include biased reports as a result of authors' own environmental views or, in the case of SPMs, because of government representatives' input (Schrope, 2001) and the tendency of IPCC reports to understate the urgency of the situation by limiting the timespan of research to the end of 2100, and by adopting conservative estimates of the changes and the risks these pose (Leuschner, 2016; Risbey, 2008). The treatment of uncertainty, both in general and across the working groups, is also criticised (Adler and Hadorn, 2014).

While such liberal critiques of IPCC practices are clearly motivated by a desire to address the need to incorporate alternative world views in climate change reports, as discussed previously many critiques of the IPCC's work are motivated by the desire to protect the interests of influential industry lobby groups. Indeed, a critique of the IPCC's work can sometimes even be motivated by diametrically opposed agendas on the part of the critics, and the issues surrounding the reporting of uncertainties in IPCC reports is a good example of this. Drawing on the example of how the assumptions of the 1970s scientific models of the processes involved in ozone depletion turned out to be oversimplified to the extent that they failed to predict the large 'ozone hole' observed over the Antarctic in the 1980s, Oppenheimer and colleagues (2007) argue that uncertainties in climate change models should be reported on in order to *promote* taking action on reducing GHG emissions. The danger that the uncertainties involved in climate change research that the IPCC assesses and reports on is being similarly downplayed leads the authors to suggest that the emphasis on consensus when finalising climate change reports is misguided, and that it is 'important that policy-makers understand the more extreme possibilities that consensus may exclude or downplay' (Oppenheimer et al., 2007, p. 1505). On the other hand, with the aim of achieving the diametrically opposite objective of *delaying* taking action on reducing fossil fuel-based GHG emissions, the representatives of fossil fuel interest groups 'try to steer the IPCC message toward emphasizing uncertainties and greenhouse gases *other* than carbon-dioxide' (Agrawala, 1998b; emphasis in original). This example points to the validity of Hajer's argument that criticisms of the IPCC for not being 'purely scientific' miss the point since, as shown throughout the discussion in this chapter, 'the essence of its organizational practice … is, indeed, a political framing in itself' (Hajer, 2012, p. 458).

Evaluation of the IPCC: it works exactly as it was intended to work

As argued above, liberal critiques of the IPCC are misplaced; rather than this institution embodying 'the long-term liberal dream of using dispassionate scientific research as a basis for transnational policy' (Agrawala and Andresen, 1999, p. 471), it was consciously established and then consciously shaped in a way that facilitates the agenda of what Agrawala and Andresen (*ibid.*) refer to as 'liberalism's enemies'. Thus 'trapped' between criticisms from climate change deniers and criticisms from political liberals, climate scientists – who are notoriously naïve when it comes to politics and try to avoid political issues (Holt, 2017; Luton, 2015) – are in an unenviable position. The psychological stress they experience as a result of their awareness of the uncompromising reactions of the physical Earth System to further GHG-induced radiative forcing is compounded by the psychological stress caused by the attacks on their personal integrity (Biddle and Leuschner, 2015; Oreskes and Conway, 2010). The psychological stress that climate scientists are subjected to, and demands to continually respond to criticisms they have already addressed, also detract from their ability to focus on their scientific work (Leuschner, 2016) – work which is, in itself, overwhelmingly complex given the multi-faceted dynamics of the Earth System.[13] As Luton (2015, p. 157) points out, however, '[t]he establishment of the IPCC was a political act that was politically motivated', and the individual scientists volunteering to work within this institution need to develop a deeper awareness of this, as do their liberal critics.

In the final analysis, and as discussed in detail in this chapter, the origins, structure and disciplining of the IPCC make it difficult to disagree with ecosocialist arguments that capitalist institutions are incapable of solving the environmental problems that capitalist social relations of production engender. As Ian Angus advises:

> The IPCC is what it is. It isn't an activist organization, and it doesn't include the full range of climate change possibilities in its reports. It produces summaries on the scientific consensus about global warming – and it is a profound commentary on how badly capitalism has damaged our world that the IPCC's conservative statements of fact constitute a powerful indictment of the capitalist system.
>
> (Angus, 2007)

International and intergovernmental institutions such as the IPCC are particularly incapable of addressing the problem of the GHG emissions that are the by-product of the *fossil fuels* underpinning the global capitalist economy. This point is further underscored by considerations of what progress has been made in reducing GHG emissions since the establishment of the UNFCCC in 1990.

The United Nations Framework Convention on Climate Change

The UNFCCC (also referred to as the 'Framework Convention') was signed at UNCED (the first Rio Earth Summit) in 1992 and entered into force in 1994 (UNFCCC, 2006). The Framework Convention constitutes the 'basic framework' of climate change governance and is 'largely procedural'; as Bodansky (2001, p. 32) explains, its 'main value is to establish a legal and institutional framework for future work through regular meetings of the parties and the possible adoption of more substantive protocols'. The UNFCCC took this form because negotiations during its drafting were contentious, with the US and allied governments refusing to countenance binding 'targets and timetables' for reducing GHG emissions (Bodansky, 2001).[14]

A very important contested issue during the initial negotiations of the UNFCCC centred around the concept of 'common but differentiated responsibilities' (CBDR), which acknowledges the historical responsibility of the advanced capitalist countries for the accumulation of most historical GHG emissions and was intended to accommodate what was seen as the 'development needs' of the Global South countries by requiring developed countries to take the lead in reducing GHG emissions (Bodansky, 2010; McGee and Steffek, 2016).[15] The developmental gap between the advanced capitalist economies and Global South countries is evident when one examines Figure 4.2 (Steffen et al., 2015), which disaggregates US and OECD socio-economic trends from socio-economic trends in Brazil, Russia, India, China and South Africa (BRICS) and other 'developing' countries in the period 1750–2010.

Liberal proponents of climate justice argue that the advanced capitalist economies should lead in reducing GHG emissions because of their greater contribution to the accumulation of GHG emissions in the atmosphere and their 'greater ability to pay for emission reductions' (McGee and Steffek, 2016, pp. 43–44). These more 'developed' countries, according to mainstream arguments supporting climate justice, also have ethical obligations to 'protect future generations from climate risk', 'share remaining emissions budget on a per-capita distribution' and 'excuse poor countries from diverting resources that would provide minimum living standards for their population' (*ibid.*, p. 44). In response to these demands for climate justice, the US has been particularly insistent on the need for *all* countries (including those in the Global South, and irrespective of historical responsibility) to limit their GHG emissions (Agrawala and Andresen, 1999). Dating back to the initial UNFCCC

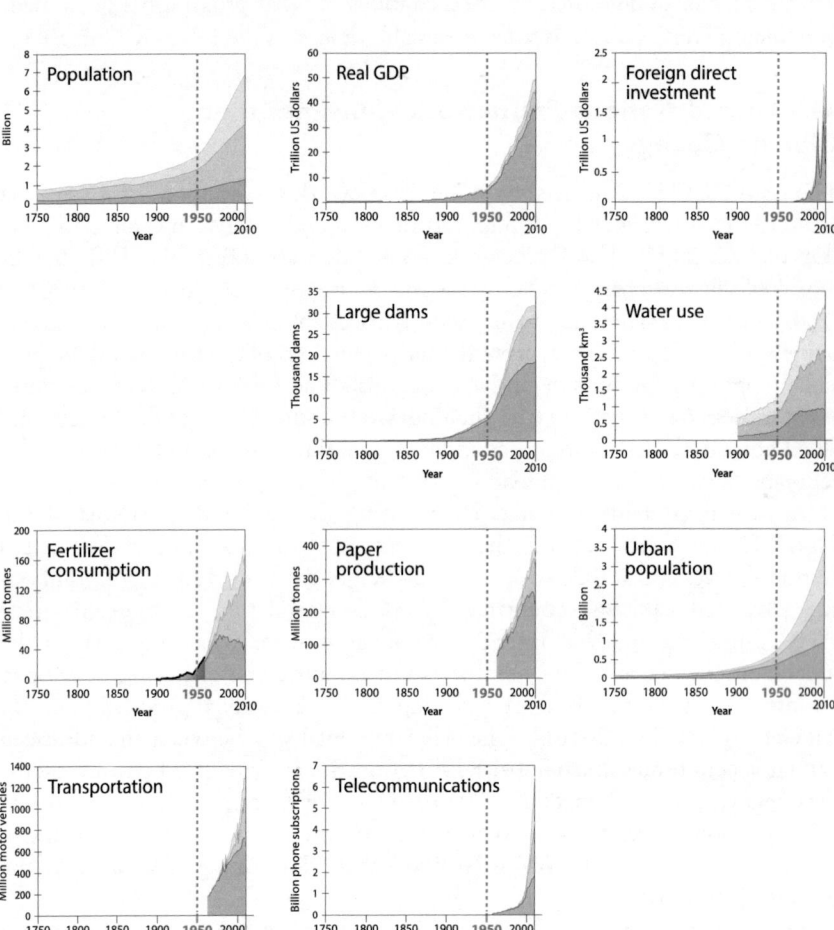

Figure 4.2   Socio-economic trends (OECD, BRICS, others).

Source: Steffen et al. (2015). © 2015 the Authors. Reprinted by permission of SAGE Publications, Ltd.[16]

negotiations, US policymakers have also insisted on the establishment of strong implementation mechanisms, detailed reporting requirements and the adoption of a noncompliance procedure (Bodansky, 2001). As discussed further in Chapter 5, accounting measures such as these are prerequisites for 'commodifying' carbon (in other words, for transforming CO_2 emissions into tradable commodities) and using 'market mechanisms' such as 'carbon trading' to create new avenues for capital accumulation. Bodansky (2001) concludes that the Framework Convention adopted in 1992 'papers over' rather than resolves the Global North/Global South divisions.

In liberal terms, then, the UNFCCC is best described as aspirational and, at the insistence of US negotiators, contains only 'ambiguous language' regarding GHG emission reduction commitments. Since the UNFCCC does not demand any binding commitments, the governments of all developed countries other than Turkey ratified it quickly (Gupta, 2010, p. 640). The George H.W. Bush Administration, too, signed the UNFCCC at the Earth Summit in 1992 and the US Senate ratified it later that year (Agrawala and Andresen, 1999). Being a party to the UNFCCC allows the US to attend meetings held under its auspices and thus to shape its evolution. It is for this reason that influential actors, including representatives of large fossil fuel corporations such as ExxonMobil, expressed opposition to the Trump Administration's decision to withdraw from the Paris Agreement (Milman, Smith and Carrington, 2017); as a ConocoPhillips spokesperson put it, being party to the Paris Agreement 'gives the U.S. the ability to participate in future climate discussions to safeguard its economic and environmental best interests' (Lui, 2017). It is for the same reason that ANU Environmental Policy academic Luke Kemp maintains that 'A US withdrawal would be the best outcome for international climate action' because 'the US and the Trump administration can do more damage inside the agreement than outside it' (Kemp, 2017a; see also Kemp, 2017b). As in the case of the IPCC, US influence on the evolution of the UNFCCC has been evident since its initial establishment.

Gupta (2010, p. 640) identifies five sets of principles enshrined in the UNFCCC: CBDR and respective capabilities; sensitivity to the needs of particularly vulnerable countries; adoption of the precautionary approach subject to cost-effectiveness; the right of all countries to pursue sustainable development; and the need to support an open international economic system.[17] The contested issues and the neoliberal economic principles enshrined in the UNFCCC have continued to block progress at the annual Conference of the Parties (COP). The first COP was held in 1995 in Berlin, and its outcome was the 'Berlin Mandate' to develop a protocol requiring developed countries to decrease their GHG emissions (Agrawala and Andresen, 1999). More than two decades later, 25 COPs have been convened, resulting in several agreements, declarations, accords, plans of action and even a protocol (the Kyoto Protocol, which came into force in 2005) – all with little (if any) effect on reducing GHG emissions. Even within the narrow confines of a liberal reformist vision, Vihma's (2011, p. 6) argument that '[t]he major challenge for the legitimacy of the UN-based climate regime has been the lack of substantive decisions, in other words its lack of effectiveness' is even more valid today than it was when it was written, given that all plans of agreeing to binding GHG emission reduction targets and timetables for achieving such reductions have now been successfully defeated by successive US administrations and their allies.

Citing US negotiators' success in removing any mention of targets and timetables in the UNFCCC, and US success in the Kyoto Protocol's inclusion of 'flexible' market mechanisms such as emissions trading, Joint Implementation (JI) and the Clean Development Mechanism (CDM), Thompson (2006) argues that the US influence over the nature of the evolving climate regime has been 'disproportionate' (see also Agrawala and Andresen, 1999). By refusing to ratify the Kyoto Protocol, initiating a shift to 'voluntary' GHG emission targets and pushing forcefully for 'developing country participation' in GHG emission reductions at COP-15, and by entrenching the so-called 'bottom-up' approach while also ensuring a very weak outcome in the form of the 'Paris Agreement' drafted at the 2015 COP-21, the US has effectively taken the lead in blocking effective climate action. As Wirth (2016, p. 169) notes, because the US Intended Nationally Determined Contributions (INDCs) in the Paris Agreement are non-binding, and because all INDCs under this agreement 'have the same international legal character', *none* of the participating countries' INDCs are legally binding.

Material capabilities: the effects of the failure of the official climate change 'regime' to reduce GHG concentrations on the biosphere

The neoliberal ideology of 'open' and 'free' global markets informing the determination of the US and its allies to maintain conditions conducive to 'business as usual' has resulted in increasing global CO_2 concentrations from 363.3 ppm since COP-1 in 1995 to 415.19 ppm in March 2020 (Scripps Institution of Oceanography, 2020). Moreover, research by Foster, Royer and Lunt (2017) shows that contemporary CO_2 concentrations are not only higher than they have been at any time in the past 200 million years, but are also increasing more rapidly than at any point in at least the past 66 million years. This is despite the efforts of the many dedicated scientists and other experts who volunteer their time to work within the IPCC, and despite a more 'enlightened' faction of capital arguing for 'green economic growth' and the activism of a large, albeit divided, climate movement that has developed within civil society to exert pressure on policymakers to take action on climate change.

Notes

1 Citing Stephen Krasner's definition of a regime as a set 'of implicit or explicit principles, norms, rules, and decision-making procedures', Zelli (2011, pp. 255–256) points out that 'a regime can be identical with a single treaty, but usually embraces a larger set of agreements under the same legal umbrella and associated policy processes', including not only treaties such as the Kyoto Protocol but also the UNFCCC and regulations of other organisations which also regulate policies relevant for climate change (for example, the WTO).

2 At least some IPCC scientists seem to be aware of this issue, as evidenced by the comments one of the IPCC authors made at a February 2017 Expert Meeting on Communications organised by the body to discuss its communication strategies for AR6. In response to a colleague who argued that the IPCC had failed in its

efforts to communicate the urgency of the situation, one of the participating scientists said: 'The mandate of the IPCC is to be relevant without being prescriptive ... This is very [restricting] ... In a sense, we are like a physician who is allowed to diagnose a sickness, to comment on a list of potential treatments, but who is prevented ... [from] prescrib[ing] a specific treatment' (IPCC, 2017).

3 This perception of the origins of the modern environmental movement is, however, inaccurate and simplistic, as discussed in Chapter 5.

4 Some of the formal but non-binding outcomes from the Rio Summits are: Agenda 21 (UNGA, 1992), the Johannesburg Declaration on Sustainable Development (UNGA, 2002), and *The Future We Want* (UNGA, 2012).

5 Zelli (2011) details how some policy proposals for addressing climate change (such as Reducing Emissions from Deforestation and Forest Degradation [REDD] and the expansion of biofuel production) are inconsistent with the regime to protect biodiversity (the Convention on Biological Diversity), while the hydrofluorocarbons (HFCs) and petrofluorocarbons (PFCs) promoted as 'safe' substances in the Vienna Convention for the Protection of the Ozone Layer are greenhouse gases that need to be phased out under the climate regime.

6 According to Agrawala (1999), policymakers soon sidelined the AGGG by refusing to fund it and generally ignoring it, and creating an alternative institution instead: the IPCC. Boehmer-Christiansen's (1994, p. 189) account of the AGGG's demise differs, claiming that this body 'was disbanded under pressure from the US State Department and presumably with the support of WMO and oil exporting countries'.

7 While China overtook the US as the largest CO_2 emitter in 2005 (Olivier et al., 2016, p. 43), the role of 'pollution havens' (whereby emissions-intensive production processes are 'offshored' from developed to developing countries such as China) substantially contributes to these emissions. Accounting methods that incorporate the consumption of CO_2 embedded in the measurement of Chinese exports reveal that the GHG emissions associated with the production of these goods have effectively been 'exported' to China by US and other multinational corporations based in the advanced capitalist economies (Aichele and Felbermayr, 2015; Malik and Lan, 2016). In addition, when measured on a per-capita basis, Chinese GHG emissions are much lower than those of many other countries, with Canadian, Australian and US per-capita emissions from fossil fuel use and cement production being the highest and China's per-capita emissions in these sectors ranking fourteenth on the list (Olivier et al., 2016, p. 40; refer also to Falkner, 2005, p. 591 for more details about per-capita US consumption patterns).

8 Refer to Ruckelshaus (1985) for an account of how the EPA was similarly established in a way that limited its power to fulfil its mandate effectively.

9 Luton (2015) presents a comprehensive argument against the claim of 'policy neutrality' within the IPCC.

10 The IPCC has published five major assessment reports to date: the First Assessment Report (FAR) in 1990, while policymakers were negotiating the UNFCCC; the Second Assessment Report (SAR) in 1995, which was used to inform the Kyoto Protocol negotiations; the Third Assessment Report (TAR) in 2001; the Fourth Assessment Report (AR4) in 2007; and the Fifth Assessment Report (AR5) in 2013/2014 (IPCC, n.d.). It plans to complete the Sixth Assessment Report (AR6) in 2021/2022 (IPCC n.d.). In addition to the primary assessment reports, the IPCC also publishes Special Reports on specific topics agreed on by member countries, with three of the most recent such reports being the 2018 Special Report on *Global Warming of 1.5 °C* and the 2019 *Special Reports on Climate*

Change and Land and *The Ocean and Cryosphere in a Changing Climate* (www.ipcc.ch/reports/).

11 There are three levels of endorsement of IPCC reports, the strongest of which is 'approval' and involves 'detailed line-by-line discussion and agreement' by government representatives while 'adoption', which is used for the Synthesis Report, is subject to 'section by section' agreement. 'Acceptance' is the weakest form of endorsement and signifies that the material 'presents a comprehensive, objective and balanced view of the subject matter' (IPCC, n.d.).

12 Connolly and Connolly (2014) explain the technical, scientific controversies over the use of proxy data in the MBH studies and Hicks (2017) provides an account of the importance of 'inductive risk' and of the need to balance 'false negatives' and 'false positives' in scientific investigations of AGW.

13 Empathy with the difficulty climate scientists face as a result of climate denier attacks, coupled with concerns about the physical ramifications of 'business-as-usual' and of the 'insidious effect on research' as climate scientists understate the extent and the effects of global climate change or try to avoid doing certain kinds of research in order to avoid harassment even leads some theorists to argue that climate scientists should consider not conducting 'dissenting' research that could be used in a way that is 'epistemically detrimental'. Biddle and Leuschner (2015) and Leuschner (2016) describe the conditions under which dissenting research could be considered to be epistemically detrimental. With reference to the controversy surrounding Bjørn Lomborg's 2001 book *The Skeptical Environmentalist*, Balint (2003) summarises some scientists' perspectives on how such work is epistemically damaging as well as potentially contributing to negative practical outcomes.

14 US allies on issues related to climate change have varied over time, depending on the issue and on how it aligns with other negotiating parties' agendas. Their most frequent allies have been oil-producing states in both the Global North and the Global South (Depledge, 2008).

15 Pre-empting being held financially liable for its share of historical emissions, US interpretations of CBDR insist that this principle does not allocate responsibility for past GHG emissions (McGee and Steffek, 2016).

16 The material reprinted in Figure 4.2 is the exclusive property of SAGE Publishing and is protected by copyright and other intellectual property laws. User may not modify, publish, transmit, participate in the transfer or sale of, reproduce, create derivative works (including course packs) from, distribute, perform, display, or in any way exploit any of the content of the file(s) in whole or in part. Permission may be sought for further use from Publications Ltd., Rights & Permissions Department, 1, Oliver's Yard, 55 City Road, London EC1Y 1SP, Email: permissions@sagepub.co.uk. By accessing the file(s), the User acknowledges and agrees to these terms. Licensor's website: www.sagepub.co.uk.

17 The notion of 'respective capabilities' refers to the different capabilities that countries have to address the problem (Gupta, 2010).

References

Adler, C.E. and Hadorn, G.H. (2014). The IPCC and Treatment of Uncertainties: Topics and Sources of Dissensus. *WIREs Climate Change*, 5, 663–676.

Agrawala, S. (1998a). Context and Early Origins of the Intergovernmental Panel on Climate Change. *Climatic Change*, 39, 605–620.

Agrawala, S. (1998b). Structural and Process History of the Intergovernmental Panel on Climate Change. *Climatic Change*, 39, 621–642.

Agrawala, S. (1999). Early Science-Policy Interactions in Climate Change: Lessons from the Advisory Group on Greenhouse Gases. *Global Environmental Change*, 9(2), 157–169.

Agrawala, S. and Andresen S. (1999). Indispensability and Indefensibility? The United States in the Climate Treaty Negotiations. *Global Governance*, 5(4), 457–482.

Åhman, M., Nilsson, L.J. and Johansson, B. (2016). Global Climate Policy and Deep Decarbonisation of Energy-Intensive Industries. *Climate Policy*, 17(5), 6340649.

Aichele, R. and Felbermayr, G. (2015). Kyoto and Carbon Leakage: An Empirical Analysis of the Carbon Content of Bilateral Trade. *The Review of Economics and Statistics*, 91(1), 104–115.

Angus, I. (2007). The IPCC and the Conservatism of Consensus. Climate & Capitalism, 5 April. Available at http://climateandcapitalism.com/2007/04/05/the-ipcc-and-the-conservatism-of-consensus.

Armitage, K.C. (2005). State of Denial: The United States and the Politics of Global Warming. *Globalizations*, 2(3), 417–427.

Baer, H.A. (2014). Activist Profile: James Hansen. In Dietz, M. and Garrelts, H. (eds), *Routledge Handbook of the Climate Change Movement*. New York: Routledge.

Baer, H.A. (2016). The Nexus of the Coal Industry and the State in Australia: Historical Dimensions and Contemporary Challenges. *Energy Policy*, 99, 194–202.

Bainbridge A. (2017). Investors Snapping Up Community Energy Projects, with Some Selling Out in Minutes. Australian Broadcasting Corporation, 29 April. Available at www.abc.net.au/news/2017-04-30/community-energy-projects-selling-out-within-minutes/8476794.

Balint, P.J. (2003). How Ethics Shape the Policy Preferences of Environmental Scientists: What We Can Learn from Lomborg and His Critics. *Politics and the Life Sciences*, 22(1), 14–23.

Beeson, M. and McDonald, M, (2013). The Politics of Climate Change in Australia. *Australian Journal of Politics and History*, 59(3), 331–348.

Biddle, J.B. and Leuschner, A. (2015). Climate Scepticism and the Manufacture of Doubt: Can Dissent in Science be Epistemically Detrimental? *European Journal for Philosophy of Science*, 5(3), 261–278.

Biermann, F. (2013). Curtain Down and Nothing Settled: Global Sustainability Governance after the 'Rio+20' Earth Summit. *Environment and Planning C: Government and Policy*, 31(6), 1099–1114.

Bodansky, D. (2001). The History of the Global Climate Change Regime. In Luterbacher, U. and Spriz, D.F. (eds), *International Relations and Global Climate Change*. Cambridge, MA: The MIT Press, 23–40.

Bodansky, D. (2010). The Copenhagen Climate Change Conference: A Postmortem. *The American Journal of International Law*, 104(2), 230–240.

Boehmer-Christiansen, S. (1994). Global Climate Protection Policy: The Limits of Scientific Advice, Part 2. *Global Environmental Change*, 4(3), 185–200.

Brundtland, G.H. (1987). Our Common Future – Call for Action. *Environmental Conservation*, 14(4), 291–294.

Committee for Skeptical Inquiry (CSI) (2015). Deniers and not Skeptics. 5 December. Available at www.csicop.org/news/show/deniers_are_not_skeptics.

Connolly, R. and Connolly, M. (2014). Global Temperature Changes of the Last Millennium. *Open Peer Review Journal* (Climate Science), 16(1), 1–50.

Corbera, E., Calvet-Mir, L., Hughes, H. and Paterson, M. (2015). Patterns of Authorship in the IPCC Working Group III Report. *Nature Climate Change*, 6(1), 94–99.

Crowley, K. (2017). Up and Down with Climate Politics 2013–2016: The Repeal of Carbon Pricing in Australia. *WIREs Climate Change*, 8, e458.

Depledge, J. (2008). Striving for No: Saudi Arabia in the Climate Change Regime. *Global Environmental Politics*, 8(4), 9–35.

DIVERSITAS and International Council for Sciences (ICSU). (2012). *State of the Planet Declaration*. Available at www.icsu.org/news-centre/news/state_of_planet_declaration.pdf.

Falkner, R. (2005). American Hegemony and the Global Environment. *International Studies Review*, 7(4), 585–599.

Fløttum, K., Gasper, D. and St Clair, A.L. (2016). Synthesizing a Policy-Relevant Perspective from the Three IPCC 'Worlds' – A Comparison of Topics and Frames in the SPMs of the Fifth Assessment Report. *Global Environmental Change*, 38, 118–129.

Folland, C.K., Karl, T.R., Christy, J.R., Clarke, R.A., Gruza, G.V., Jouzel, J., Mann, M.E., Oerlemans, J., Salinger, M.J. and Wang, S.-W. (2001). Observed Climate Variability and Change. In Houghton, J.T., Ding, Y., Griggs, D.J., Noguer, M., van der Linden, P.J., Dai, X., Maskell, K. and Johnson, C.A. (eds), *Climate Change 2001: The Scientific Basis. Contribution of Working Group I to the Third Assessment Report of the Intergovernmental Panel on Climate Change*. Cambridge and New York: Cambridge University Press, 99–181.

Foster, G., Royer, D. and Lunt, D. (2017). We Are Heading for the Warmest Climate in Half a Billion Years, Says New Study. The Conversation, 4 April. Available at https://theconversation.com/we-are-heading-for-the-warmest-climate-in-half-a-billion-years-says-new-study-73648.

Gupta, J. (2010). A History of International Climate Change Policy. *WIREs Climate Change*, 1, 636–653.

Hadden, J. and Seybert, L.A. (2016). What's In a Norm? Mapping the Norm Definition Process in the Debate on Sustainable Development. *Global Governance*, 22, 249–268.

Hajer, M.A. (2012). A Media Storm in the World Risk Society: Enacting Scientific Authority in the IPCC Controversy (2009–10). *Critical Policy Studies*, 6(4), 452–464.

Hamblyn, R. (2009). The Whistleblower and the Canary: Rhetorical Constructions of Climate Change. *Journal of Historical Geography*, 35(2), 223–236.

Harris, P.G. (2009). Beyond Bush: Environmental Politics and Prospects for US Climate Policy. *Energy Policy*, 37(3), 966–971.

Hecht, A.D. and Tirpak, D. (1995). Framework Agreement on Climate Change: A Scientific and Policy History. *Climatic Change*, 29, 371–402.

Hicks, D.J. (2017). Scientific Controversies as Proxy Politics. *Issues in Science and Technology*, 33(2), 67–72.

Hulme, M. and Mahony, M. (2010). Climate Change: What Do We Know about the IPCC? *Progress in Physical Geography*, 34(5), 705–718.

InterAcademy Council (IAC) (2010). *Climate Change Assessments: Review of the Processes and Procedures of the IPCC*. Available at www.interacademycouncil.net/24026/26050.aspx.

Intergovernmental Panel on Climate Change (IPCC). (2010). *Understanding Climate Change: 22 Years of IPCC Assessment*. Available at http://ipcc.ch/pdf/press/ipcc_leaflets_2010/ipcc-brochure_understanding.pdf.

Intergovernmental Panel on Climate Change (IPCC). (2017). IPCC Expert Meeting on Communications, Day 2 – Recommendations, Conclusions and Explanation of Next Steps. Available at www.youtube.com/watch?v=m1Nc-eGQMvQ.

Intergovernmental Panel on Climate Change (IPCC) (n.d.). *Appendix A to the Principles Governing IPCC Work*. Available at www.ipcc.ch/site/assets/uploads/2018/09/ipcc-principles-appendix-a-final.pdf.

International Council for Science (ICSU) (2015). *The International Council for Science and Climate Change, 60 Years of Facilitated Climate Change Research and Informing Policy*. Available at https://council.science/wp-content/uploads/2017/04/ICSU_and_Climatechange.pdf.

Kemp, L. (2017a). The World would be Better Off if Trump Withdraws from the Paris Climate Deal. The Conversation, 23 May. Available at https://theconversation.com/the-world-would-be-better-off-if-trump-withdraws-from-the-paris-climate-deal-78096.

Kemp, L. (2017b). US-Proofing the Paris Climate Agreement. *Climate Policy*, 17(1), 86–101.

Kovel, J. (2007). *The Enemy of Nature: The End of Capitalism or the End of the World?* New York: Zed Books.

Leuschner, A. (2016). Is it Appropriate to 'Target' Inappropriate Dissent on the Normative Consequences of Climate Skepticism? *Synthese*, 195, 1255–1271.

Longo, S.B., Clausen, R. and Clark, B. (2015). *The Tragedy of the Commodity: Oceans, Fisheries, and Aquaculture*. New Brunswick, NJ: Rutgers University Press.

Lui, K. (2017). Top CEOs are in a Last Ditch Bid to Persuade Trump to Stick with the Paris Climate Deal. *Fortune*, 1 June. Available at http://fortune.com/2017/06/01/apple-google-facebook-trump-stay-paris-accord/.

Luton, L.S. (2015). Climate Scientists and the Intergovernmental Panel on Climate Change: Evolving Dynamics of a Belief in Political Neutrality. *Administrative Theory & Praxis*, 37(3), 144–161.

Malik, A. and Lan, J. (2016). The Role of Outsourcing in Driving Global Carbon Emissions. *Economic Systems Research*, 28(2), 168–182.

Mann, M.E., Bradley, R.S. and Hughes, M.K. (1998). Global-Scale Temperature Patterns and Climate Forcing over the Past Six Centuries. *Nature*, 392, 779–787.

McAdam, D. (2017). Social Movement Theory and the Prospects for Climate Change Activism in the United States. *Annual Review of Political Science*, 20, 189–208.

McGee, J. and Steffek, J. (2016). The Copenhagen Turn in Global Climate Governance and the Contentious History of Differentiation in International Law. *Journal of Environmental Law*, 28, 37–63.

Meadows, D., Randers, J. and Meadows, D. (2004). *Limits to Growth: The 30-Year Update*. White River Junction, VT: Chelsea Green Publishing Company.

Milman, O. (2015). James Hansen, Father of Climate Change Awareness, Calls Paris Talks 'A Fraud'. *The Guardian*, 12 December. Available at www.theguardian.com/environment/2015/dec/12/james-hansen-climate-change-paris-talks-fraud.

Milman, O., Smith, D. and Carrington, D. (2017). Donald Trump Confirms US Will Quit Paris Climate Agreement. *The Guardian*, 1 June. Available at www.theguardian.com/ environment/2017/jun/01/donald-trump-confirms-us-will-quit-paris-climate-deal.

Morgan, W. (2017). Coal Comfort: Pacific Islands on Collision Course with Australia over Emissions. The Conversation, 28 February. Available at https://theconversation. com/coal-comfort-pacific-islands-on-collision-course-with-australia-over-emissions-73662.

Obermeister, N. (2017). From Dichotomy to Duality: Addressing Interdisciplinary Epistemological Barriers to Inclusive Knowledge Governance in Global Environmental Assessments. *Environmental Science & Policy*, 68, 80–86.

Olivier, J.G.J., Janssens-Maenhout, G., Muntean, M. and Peters, J.A.H.W. (2016). *Trends in Global CO2 Emissions, 2016 Report*. The Hague: PBL Netherlands Environmental Assessment Agency; Ispra: European Commission, Joint Research Centre.

Oppenheimer, M., O'Neill, B.C., Webster, M. and Agrawala, S. (2007). The Limits of Consensus. *Science*, 317(5844), 1105–1106.

Oreskes, N. and Conway, E.M. (2010). *Merchants of Doubt*. New York: Bloomsbury Press.

Pelley, J. (2015). Canada Reclaims its Scientific Heritage. *Frontiers in Ecology and the Environment*, 13(10), 528.

Peterson, T.C., Connolley, W.M. and Fleck, J. (2008). The Myth of the 1970s Global Cooling Scientific Consensus. *Bulletin of the American Meteorological Society*, 89(9), 1325–1337.

Ravindranath, N.H. (2010). IPCC: Accomplishments, Controversies and Challenges. *Current Science*, 99(1), 26–35.

Rich, R.F. and Merrick, K.R. (2007). Use and Misuse of Science: Global Climate Change and the Bush Administration. *Virginia Journal of Social Policy & the Law*, 13(3), 223–252.

Risbey, J.S. (2008). The New Climate Discourse: Alarmist or Alarming? *Global Environmental Change*, 18(1), 26–37.

Rockström, J., Steffen, W., Noone, K., Persson, Å., Chapin, F.S., III, Lambin, E., Lenton, T.M., Scheffer, M., Folke, C., Schellnhuber, H.J., Nykvist, B., de Wit, C.A., Hughes, T., van der Leeuw, S., Rodhe, H., Sörlin, S., Snyder, P.K., Costanza, R., Svedin, U., Falkenmark, M., Karlberg, L., Corell, R.W., Fabry, V.J., Hansen, J., Walker, B., Liverman, D., Richardson, K., Crutzen, P. and Foley, J. (2009). Planetary Boundaries: Exploring the Safe Operating Space for Humanity. *Ecology and Society*, 14(2), 32.

Ruckelshaus, W.D. (1985). Environmental Protection: A Brief History of the Environmental Movement in America and the Implications Abroad. *Environmental Law*, 15(3), 455–469.

Sale, K. (1993). *The Green Revolution: The American Environmental Movement, 1962–1992*. New York: Hill and Wang.

Schrope, M. (2001). Consensus Science, or Consensus Politics? *Nature*, 412, 112–114.

Scripps Institution of Oceanography (2020). The Keeling Curve. Available from https://scripps.ucsd.edu/programs/keelingcurve/.

Skrydstrup, M. (2013). Tricked or Troubled Natures? How to Make Sense of 'Climategate'. *Environmental Science & Policy*, 28, 92–99.

Steffen, W., Broadgate, W., Deutsch, L., Gaffney, O. and Ludwig, C. (2015). The Trajectory of the Anthropocene: The Great Acceleration. *The Anthropocene Review*, 2(1), 81–98.

Swann, T. (2016). Leaving the ARENA: Fossil Fuels vs Renewables in Australian Energy R&D Funding. The Australia Institute, Discussion Paper, September 2016. Available at www.tai.org.au/content/leaving-arena.

Tokar, B. (2014). *Toward Climate Justice: Perspectives on the Climate Crisis and Social Change*. Porsgrunn, Norway: New Compass Press.

Thompson, A. (2006). Management under Anarchy: The International Politics of Climate Change. *Climatic Change*, 78(1), 7–29.

Union of Concerned Scientists (UCS) (2012). A Climate of Corporate Control: How Corporations have Influenced the U.S. Dialogue on Climate Science and Policy. 31 May. Available from www.ucsusa.org/our-work/center-science-and-democracy/fighting-misinformation/a-climate-of-corporate-control.html.

United Nations Framework Convention on Climate Change (UNFCCC) (2006). *United Nations Framework Convention on Climate Change Handbook*. Available at http://unfccc.int/resource/docs/publications/handbook.pdf.

United Nations General Assembly (UNGA) (1972). *Report of the United Nations Conference on the Human Environment*. Available at www.un-documents.net/aconf48-14r1.pdf.

United Nations General Assembly (UNGA) (1992). Agenda 21. Available at https://sustainabledevelopment.un.org/content/documents/Agenda21.pdf.

United Nations General Assembly (UNGA) (2002). Johannesburg Declaration on Sustainable Development. Available at www.un-documents.net/jburgdec.htm.

United Nations General Assembly (UNGA) (2012). *The Future We Want*. Available at http://icriforum.org/sites/default/files/UNGA_the_future_we_want.pdf.

Vihma, A. (2011). *A Climate of Consensus: The UNFCCC Faces Challenges of Legitimacy and Effectiveness*. Briefing Paper 75, March 2011, The Finnish Institute of International Affairs. Available at www.files.ethz.ch/isn/127655/UPI_Briefing_Paper_75.pdf.

Williams, C. (2010). *Ecology and Socialism: Solutions to Capitalist Ecological Crisis*. Chicago, IL: Haymarket Books.

Wirth, D.A. (2016). Cracking the American Climate Negotiators' Hidden Code: United States Law and the Paris Agreement. *Climate Law*, 6(1), 152–170.

World Bank (2014). *Turn Down the Heat: Confronting the New Climate Normal*. Available at www-wds.worldbank.org/external/default/WDSContentServer/WDSP/IB/2014/11/20/000406484_20141120090713/Rendered/PDF/927040v20WP00O0ull0Report000English.pdf.

World Commission on Environment and Development (WCED) (1987). *Our Common Future*. Available from www.un-documents.net/our-common-future.pdf.

Zelli, F. (2011). The Fragmentation of the Global Climate Governance Architecture. *WIREs Climate Change*, 2(2), 255–270.

Zillman, J.W. (2009). A History of Climate Activities. *WMO Bulletin*, 58(3), 141–150.

5 The social dynamics of 'climate justice' versus 'climate action' in the climate movement

Introduction

Given the inadequacy of official responses to the anthropogenic climate crisis discussed in the previous chapter, concerned individuals and social movement actors have been demanding effective action from policymakers by participating in what can broadly be conceived of as a 'climate movement'. This chapter explores the origins, evolution and effectiveness of the climate movement. Using the modified 'Method of Historial Structures' (MHS) schema outlined in Chapter 2, the analysis presented in this chapter incorporates discussions of the ways in which the social dynamics of the climate movement relate to formal institutional responses to climate change and how these play out at global, state and local levels. It also incorporates some ways in which different factions of capital and labour (Cox's 'social forces') impact on, and are influenced by, the social dynamics unfolding within the climate movement, and is furthermore conducted with reference to material capabilities, dominant institutions, social facts and competing ideas.

Social facts informing individual responses to climate change

As discussed in the previous chapter, in the years following Hansen's 1988 testimony, an increasing number of scientific reports have been published emphasising, with ever greater urgency, the need to take swift and effective action to mitigate anthropogenic global warming (AGW). Much of the scientific evidence of this warming and its effects is summarised in the IPCC's Assessment Reports, and the widespread publicity accompanying the release of these reports attracts attention from concerned individuals and groups within civil society. In addition, as weather patterns change because of AGW, 'extreme weather' events increase in duration and intensity around the world (Steffen et al., 2017; WMO, 2016) and many communities are directly impacted when their lives are disrupted by unusually extensive and intense floods, droughts and wildfires, making the effects of AGW more difficult to ignore.

The apparent incongruity between the knowledge available about the causes of AGW and the lack of interest and response to this knowledge has prompted an increasing number of investigations trying to identify possible barriers at play (for example, Adams, 2017; Blühdorn, 2017; Hausknost, 2017). Given the prevailing ideology that promotes individualism in capitalist societies, it is not surprising that many of these investigations focus on individuals' psychological dispositions rather than on the systemic socio-economic and political decisions that cause AGW. Studies focusing on cognitive barriers to climate action are motivated by attempts to frame communications about climate change in ways that will prompt individuals to change their behaviour so that it is 'sustainable' (Spence, Poortinga and Pidgeon, 2012). The reality is, however, that individuals are embedded within larger social and institutional contexts that both shape the individualistic modes of thinking and simultaneously limit the effectiveness of any individual's attempts to achieve a more sustainable lifestyle. Rather than limiting efforts to attempting to change the way individual people think and behave, an understanding of these broader socio-economic and political contexts is crucial if we are to effectively address the challenges we face.

Limiting the rise in average global temperatures to 1.5 °C, or even to what some scientists see as a very dangerous 2 °C, will at the very least require rapid large-scale and comprehensive transformations of the global energy system as well as of energy use related to the transportation, production and construction sectors (Rogelj et al., 2015). It will also involve implementing policies to facilitate the large-scale removal of CO_2 from the atmosphere (*ibid.*). These are not things that individuals have either the resources or the authority to do, no matter how committed they are to living sustainably. Discussing the way in which 'city people' have no understanding of what life is like for poor people living in rural areas in the US, one of Cramer's (2017) interviewees gives an example of how exhortations by rich liberal critics not to 'drive as much' are meaningless: 'You gotta drive 20 miles to work? You can't cut that in half.' This is just one example of how realistic solutions for reducing CO_2 emissions involve large, systemic changes that transcend individual values and choices; in this case, people either have to be able to make a living or sustain themselves close to the places where they live, or they must have access to affordable and effective public transport systems to commute to work. Yet, as journalist Martin Lukacs (2017) notes in his aptly titled article 'Neoliberalism Has Conned Us into Fighting Climate Change as Individuals', 'corporate ads, school textbooks, and the campaigns of mainstream environmental groups, especially in the west' exhort individuals to change their lightbulbs, buy 'eco-appliances' and install solar panels while the GHG emissions and widespread environmental damage caused by the normal operations of global capitalism render these individual efforts irrelevant.

While individuals cannot address the causes of the global warming crisis by changing their personal behaviour given that the scale of the changes far surpasses what can be achieved at the individual level, what they *can* do is work collectively with others in order to change the *economic, political, and social systems and institutions* that are responsible for the current interrelated environmental, economic, political and social crises. It is therefore the barriers to this sort of action that are important to identify and counter. The balance of prevailing social forces (particularly with respect to the strength and influence of the fossil fuel industry and the weakness of organised labour and civil society), the dogged determination to further the neoliberalising project of extending and intensifying capitalist relations of production and the dominance of social facts that favour individualistic, capitalist economic values all present formidable barriers in this respect.

Social forces promoting ineffective individualistic responses to climate change

Research suggests that personal psychological dispositions such as those identified above are often constructed and/or compounded by successful climate change denialist strategies designed to delay the adoption of measures that will adversely affect their financial interests (McCright et al., 2016). Climate change denialist narratives work together with conservative political beliefs and the hegemonic and politically charged neoliberalising discourse that regulating capital will impact negatively on 'economic growth' and hence on jobs (Hornsey et al., 2016). Interestingly, however, some research findings suggest that conservatives' beliefs and attitudes about climate change in the US are frequently less motivated by economic concerns than by a more general antagonism to environmentalists, who are perceived as 'green on the outside, red on the inside' (Hoffarth and Hodson, 2016). Conversely, a meta-analysis of studies examining determinants of belief in climate change finds that while there are links between political ideology and climate change beliefs, these beliefs are 'more aligned to specific identification with political parties than to underlying political ideologies' (Hornsey et al., 2016, p. 622).

In addition to ideological motivations for the popularity of climate change denialism in some countries (notably in the United States, but also in Australia), widespread economic insecurities amongst working people in the advanced capitalist economies since the 1980s also play a role in the issue of climate change not being prioritised by many people. These economic insecurities have become even more pronounced in the aftermath of the 2007/2008 Global Financial Crisis, as discussed in more detail in Chapter 6. It is not surprising that people living in precarious circumstances prioritise immediate concerns such as jobs, housing and career prospects at the expense of future possible catastrophes, and that they find arguments that dealing with climate change will damage the economy and make it more difficult to find work convincing.

However, even people who claim to be both very knowledgeable and very concerned about climate change are often apathetic when it comes to actively calling for policymakers to take effective measures to reduce anthropogenic GHG emissions. Doherty and Webler (2016) identify several possible reasons for this, including that many people do not believe that their actions can make a difference and that 'similar others' (their peers) are not taking action either. Despite the general lack of active public engagement in formal political processes, there are nevertheless some people who are not only concerned about the ineffectiveness of official responses to climate change and the many other current crises but also try to engage with these issues in a variety of ways, including by participating in the heterogeneous climate movement that emerged in the late 1980s. Some sections of the climate movement are comprised of elements of the environmental movement, which started paying attention to AGW from the early days of its emergence as an issue of global concern. The brief overview of the origins and characteristics of the environmental movement that follows provides a broader context for understanding the overtly political disagreements that have characterised the evolution of the climate movement over the more than two decades of failed official climate change negotiations.

A neo-Gramscian analysis of the development of the modern environmental movement in the advanced capitalist economies

The modern environmental movement in the advanced capitalist economies dates back to at least the 1960s, with its origins attributed to a growing public awareness of issues that extended beyond the dominant concerns of the earliest (pre-World War II) 'nature preservationist' and 'nature conservationist' environmental thinkers and organisations (Sale, 1993). While many writers identify the 1962 publication of Rachel Carson's *Silent Spring*, which provides scientific evidence of the damaging effects of pesticides on both natural ecosystems and human health, as marking the beginning of this new environmental movement, other analysts question the validity of doing this. Meyer and Rohlinger (2012) caution against such oversimplified 'big book myths', which they describe as one version of 'immaculate conception' stories used by both social movement actors and academics studying them, to explain the origins of social movements. Not only are such mythical accounts historically inaccurate, but they are also counterproductive in that they ignore the wider context within which social movements arise, flourish and succeed or fail. Moreover, these simplified accounts of the origins of social movements promote the false perception that the power of ideas arising from within civil society alone can effect social change. They edit out crucial facts such as the time and effort required to build effective social movements and the wider context of the role of prevailing material conditions and ideologies in shaping the emergence and evolution of these movements. In neo-Gramscian terms, simple stories with linear timelines and a single, identifiable 'origin' fail to take into account the complex interactions between the prevailing world order, forms of state, material capabilities, institutions, social facts and the social forces shaping the emergence and evolution of social movements.

American hegemony, welfare-statism and the emergence
of the modern environmental movement in the 1960s

In their critique of 'big book myth' accounts of the rise of environmentalism and other social movements in the United States in the 1960s, Meyer and Rohlinger (2012) highlight evidence that the US federal government had started addressing some of the issues that the new social movements were concerned about even prior to the movements' emergence. For example, they point out that the US government had commenced Congressional hearings about the pollution caused by pesticides in the 1950s (*ibid.*). Rather than sparking the beginning of the environmental movement, they argue that *Silent Spring* 'articulated and amplified a pre-existing concern', and that the US government's environmental protection measures, having begun prior to the book's publication, continued after it was published, 'but not directly in response to Carson's text' (Meyer and Rohlinger, 2012, p. 143).

When evaluating its successes, it is also significant to be mindful that the modern environmental movement developed at a time when the advanced capitalist countries were organised as liberal welfare states, and that the prevailing ideology (that constituted the 'social facts' of the time) was that governments had some responsibility for protecting 'public goods' (Meyer and Rohlinger, 2012). Whatever 'wins' the environmental movement may have achieved in pushing government institutions to legislate for environmental protection should therefore be understood in the context that national governments were not as ideologically opposed to introducing such legislation in the late 1960s and early 1970s as they are today (Kraft, 2000). While '*activist government was a critical factor in spurring social mobilization* in the history of the movements of the 1960s', support for limited government intervention in 'public affairs' has grown 'tremendously' from the beginning of the 1970s (Meyer and Rohlinger, 2012, p. 148; emphasis in original). In short, it is much more difficult to persuade governments to regulate for environmental protection in the current era of neoliberalising global capitalism.

While Meyer and Rohlinger do not discuss *why* the US government was so concerned about environmental issues in the 1950s and 1960s, Robertson (2008) focuses on this question, as well as on how the wider context of 'American Empire' (or *Pax Americana*, in neo-Gramscian Robert Cox's terms) shaped the form of the US environmental movement that emerged at this time. Robertson (2008) points out that the modern environmental movement emerged in the context of 'American Empire', when US policymakers increasingly began to link the global management of natural resources with issues of what they defined as 'US national security'. As Robertson argues:

> To fully understand how and why Americans came to look at wilderness and other aspects of the American Earth in new and more comprehensive

ways, we must look more comprehensively at how *they began to see the whole planet – the Earth itself – as in some ways American.*

(Robertson, 2008, p. 584; emphasis added)

Robertson (2008) begins his analysis of the rise of the modern environmental movement by referring to conditions leading to the rise of earlier forms of environmentalism in the Global North. Citing studies that link the rise of this earlier environmental movement to the colonial expansion commencing in the seventeenth century (which culminated, in Cox's analysis, with *Pax Brittanica*), Robertson recounts arguments that environmentalism constituted a form of 'green imperialism' during this period.

Two key points emerging from Robertson's analyses of 'green imperialism' are that 'economic and political dominance on a global scale required a degree of planning that helped promote conservationism' and that 'European conservation actually emerged from the [natural resource] *management requirements* of colonial empires' (Robertson, 2008, p. 563; emphasis added; see also Dalby, 2004). The importance of developing such an understanding of the rise of environmentalism is that it focuses attention on how environmental concerns 'often emerge from and reinforce hierarchies of power' and, relatedly, that while environmental concerns can serve as issues around which to mobilises anti-imperialist forces, they can also be used 'as a handmaiden to empire, providing imperial officials with another way to regulate and control far-off lands and peoples' (Robertson, 2008, pp. 563–564).

Following the logic of these earlier accounts of the relationship between conservationism and European colonialism, Robertson (2008, p. 564) develops an argument that growing US global power in the post-World War II years similarly 'created new imperatives to manage resources, new sciences with which to do so, new forms of environmental crises, new anti-modern doubts, and ultimately new policy frameworks', and that these need to be taken into account in explanations of why the modern environmental movement 'exploded' on the scene when it did, and why it took the forms it took. An illustrative concrete example supporting the argument of the emerging links between access to natural resources and US national security is the 'list of sixty strategic resources the United States needed, of which thirty came entirely from overseas' compiled by US government planners, who then 'took measures to guarantee their supply' (Robertson, 2008, p. 568). Conca (2004, p. 14) notes that the project to secure US access to global natural resources has succeeded, identifying the 1980s debt crisis as having been instrumental in 'lock[ing] in steady resource supplies at favourable prices' and the 1990s 'trade liberalization initiatives' as further deepening US capacity to access these resources. It should not be surprising, then, that in the context of climate change negotiations, Global South fears that the issue of global warming is being used as an excuse by powerful Global North governments to control their resources and to prevent their economic development has caused much distrust in the UNFCCC negotiations (Gupta, 2010).

Western environmental NGOs take up the issue of global warming and climate change

The growing publicity about AGW in the 1980s prompted some environmental groups in the advanced capitalist economies to turn their attention towards the end of that decade and to start collaborating in their attempts to engage with the formal institutions established to address it (Agrawala, 1999; Betsill, 2002). The current climate movement can thus be seen as having its 'pre-history' in the 1980s, with major US environmental non-governmental organisations (ENGOs) and a new coalition of NGOs, the Climate Action Network (CAN), attending the UN's climate change conferences. The ENGOs that attended the early discussions about climate change saw it as a supplementary issue to their other environmental campaigns, and it is only more recently that a distinct, albeit highly heterogeneous, 'climate movement' has emerged which is divided along at least four lines: nationality; types of actors; age and gender; and ideology (Garrelts and Dietz, 2014). Understanding the location and role of ecosocialists within this movement entails understanding its composition and identifying the lines along which divisions within it occur. This is not an easy task as the climate movement consists of a bewilderingly diverse, dynamic and fluid variety of actors that include individuals, civil society organisations (CSOs), social movement organisations (SMOs) and networks that often overlap and defy categorisation into separate, strictly defined and stable groupings.

Given the diversity of climate movement actors, it is not surprising that analysts have traced its roots to a variety of other SMOs, such as the pre-existing environmental movement, the environmental justice movement and the Global Justice Movement (GJM) in the Global North, as well as to indigenous peoples and small-scale subsistence farmers and fisherpeople (predominantly, but not exclusively, in the Global South) who are trying to protect their traditional lands and sources of reproduction that are being threatened as global capitalist relations of production expand. In view of the insurmountable challenges faced by those attempting to map the networks of component actors within the climate movement in detail, many analysts find it useful to at least distinguish between its two major 'wings' or 'streams': the more moderate wing (which is referred to as the 'climate action' wing in this book) working towards reforming global capitalism by calling for the decarbonisation of the global economy, and the system-critical 'radical climate justice' wing that sees climate change as one of many manifestations of a much wider crisis that can only be resolved by fundamental system change.[1] An awareness of the differences between these two extremes and, more importantly, of the nuanced differences in the common ground they seemingly share is crucial if one is to understand the role of ecosocialist theorists and activists within the broader climate movement. Also important in understanding the role of ecosocialists within the climate movement is being aware of the very messy 'in-between' space which many rank-and-file movement actors (as opposed to the leadership) occupy (Wahlström, Wennerhag, and Rootes, 2013) and which represents a continuum between the two extreme wings. Given these prerequisites, it is argued that the ideological division is the most crucial of the four lines of division identified by Garrelts and Dietz (2014). Despite participating in the same mass-protest actions such as the 2014 and 2015 People's Climate Marches, in this book the reformist climate action and the radical climate justice extremes are treated as *analytically* distinct and fundamentally *ideologically opposed* wings of the climate movement.[2] The ideological distinction is emphasised because it provides a useful analytical tool for discussing the key debates within the climate movement, where *climate action* activists and *radical climate justice* activists (with the latter including ecosocialists) are engaged in what Gramsci refers to as a 'war of position' in contesting their ideas within the terrain of the political arena of 'civil society'.[3]

The 'war of position' within the climate movement

Climate change is a particularly politicised issue because it 'is embedded in complex societal conflicts' which 'are rooted in *divergent interests in society*' (Bedall and Görg, 2014, p. 44; emphasis in original). The war of position between reformist and critical positions is thus largely fought through the discursive debates vis-à-vis the causes of (and hence possible solutions to) climate change between different actors within the broad climate movement (de Lucia, 2014). One way of understanding this ideological difference between the moderate and radical wings of the climate movement is to refer to neo-Gramscian concepts that distinguish between problem-solving and critical approaches outlined in Chapter 2.

The moderate climate action movement identifies the cause of climate change in largely scientific and politically 'sanitised' technical terms; for example, it focuses almost exclusively on how increases in GHG emissions in the atmosphere are changing the Earth's energy balance (Bedall and Görg, 2014; de Lucia, 2014; Tokar, 2014). This approach treats climate change as a specific, discrete and often technical 'silo issue' and leads to the conclusion that it can be addressed in isolation by 'problem-solving' and tweaking global capitalism so that it is no longer powered by fossil fuels. This leads climate action movement supporters to try to solve the problem of climate change by lobbying governments and policymakers to implement reforms to the current system. The solutions they propose include the adoption of renewable technologies for energy production and the development and implementation of a mix of regulatory policies closely aligned with 'market instruments' to achieve a transition from a global capitalist economy based on the use of fossil fuels to a 'green' global capitalist economy based on renewable energy. Many supporters of climate action believe that taking action on climate change is 'compatible with continuing economic growth' and are generally committed 'to a paradigm of *ecological modernization* according to which environmental problems can be resolved politically, economically, and technologically within the context of real existing institutions and power structures' (Garrelts and Dietz, 2014, p. 2).

While agreeing with the moderate climate action movement's identification of the physical mechanisms and material manifestations of anthropogenic global warming, the radical climate justice movement goes much further in its analyses of both the nature of the problem and its causes and consequences, pointing to relations of domination in all their forms (along the lines of class, sex, race and human attempts to dominate nature) as the primary causes of a variety of interconnected environmental, economic and socio-political problems (Bedall and Görg, 2014; Giacomini and Turner, 2015). Actors within the radical climate justice movement thus use a critical perspective that leads to a very complex understanding of climate change as only one challenge to humanity that is evolving within the context of an ever-expanding global capitalist system in which a small global elite exploits both nature and people in order to increase its economic wealth and political power (Burgmann and Baer, 2012; Tanuro, 2013). From the perspective of radical climate justice movement actors, climate change is thus only one *symptom* (albeit an extremely important and pressing symptom) of the widespread ecological destruction and social harms that result from the pervasive relations of domination that characterise global capitalism (Bedall and Görg, 2014; Steger, Goodman and Wilson, 2013). Given their analyses of the nature and causes of the multiple crises humanity currently faces, radical climate justice movement actors argue that real solutions to these interconnected problems require fundamental 'system change'.[4]

A comparison of how climate action supporters and how radical climate justice activists interpret phrases such as 'connecting the dots' demonstrates the different positions adopted by the two wings of the climate movement. The 2012 climate action movement's 350.org-initiated campaign aimed to 'connect the dots' between climate change and the increasing number of 'extreme' weather events around the world (McKibben, 2012), thereby narrowly focusing on the physical and technical aspects of anthropogenic GHG emissions and the scientific and technical aspects of their physical effects. Climate justice activists argue that a genuine understanding of the global warming crisis entails 'connecting the dots' between: the scientific evidence of anthropogenic climate change and other anthropogenic threats to the Earth's biosphere; the social, economic and political systems created by humans (and therefore subject to being changed by humans) causing the current crises; and the ethical implications of different courses of action. Ecosocialist analyses and arguments are well equipped to make strong contributions to joining the 'dots' in this war of position, thereby building on prevailing 'common sense' understandings (which can be partial, incomplete or somewhat incoherent) of the causes and dangers of AGW in order to promote what Gramsci refers to as a more coherent and developed 'good sense' of these causes and, very importantly, of the implications of different proposed solutions for different groups of people. In contrast to moderate climate action movement concerns about the physical and technical aspects of climate change, ecosocialist analyses draw attention to the ethical dimensions of the issue.

By consistently linking climate change and social justice issues, ecosocialist analyses emphasise that there is much at stake in who succeeds in winning this ideological debate; it is not overstating the case that the lives of many people depend on the outcome of this war of position and its success in building a powerful counter-hegemonic bloc to shift the debate within the climate movement from what radical climate justice activists see as dangerous, narrow problem-solving 'false solutions' and instead turn its energies towards working on 'system change' – creating a more truly sustainable society with different values.[5] The division between problem-solving and critical approaches within the larger climate movement is exemplified by its two largest umbrella organisations, Climate Action Network International (CAN-I) and *CJN!*, which are networks of a variety of NGOs, CSOs and social movements. Ecosocialists align themselves with *CJN!*, and were involved in forming the new coalition when it became clear that CAN-I had started supporting what radical climate justice advocates saw as 'false solutions'.

The concept of 'false solutions' represents a defining difference between the moderate climate action movement and the radical climate justice movement. While, on a superficial level, it is possible to discern 'averting climate change' as a broad common goal of all climate movement actors, unlike the reformist climate action movement, the radical climate justice movement is not prepared to achieve this aim using means that will put even greater burdens on the working class and on subaltern social groups that are already disadvantaged in capitalist societies. Climate justice activist publications such as *Hoodwinked in the Hothouse: False Solutions to Climate Change* (RTNA and CTW, n.d.), *The COP19 Guide to Corporate Lobbying: Climate Crooks and the Polish Government's partners in Crime* (Tansey, 2013) and the *Indigenous Peoples' Guide: False Solutions to Climate Change* (Land is Life et al., n.d.) classify as false a range of market-based solutions, such as carbon trading and 'flexible mechanisms' like the CDM and REDD+ projects. Solutions that depend on the development and implementation of sophisticated and expensive technology (such as 'clean coal' power plants, carbon capture and storage, nuclear energy, 'natural gas', waste incinerators and 'biogenic fuels' such as biomass) are also perceived to be false, as are solutions that will result in displacing populations and destroying food production in order to produce agrofuels or build megadams for the production of 'clean energy'. As de Lucia summarises, 'False solutions are generally seen to imply also a high level of centralization, economic, political, structural and technical' (de Lucia, 2014, p. 80). It is for these reasons that radical climate justice activists do not support moderate climate movement actors' implicit or explicit support of policies that aim to further privatise the commons and thereby facilitate the exploitation and dispossession of powerless people.

Climate Action Network International (CAN-I)

Widely acknowledged as the oldest and most popular actor in the climate movement (Garrelts, 2014; Guldbrandsen and Andresen, 2004), CAN-I describes its mission as being 'to promote government and individual action to limit human-induced climate change to ecologically sustainable levels' (CAN-I, n.d). Originally constituted as the Climate Action Network (CAN) and renamed Climate Action Network International in 2004 (CAN, 2014), this umbrella organisation represents the reformist wing of the climate movement and was established as a coalition of NGOs in 1989 'with the central objective of ensuring the 1992 UN Conference on the Environment and Development (UNCED) would implement strong emissions reductions' (Steger, Goodman and Wilson, 2013, p. 137). In addition to its 'insider' tactics of working with officials in the COPs, CAN-I members also engage in 'outsider' tactics like protest marches and other publicity-raising campaign events.

Bedall and Görg (2014, p. 49) argue that after the adoption of the Kyoto Protocol, CAN-I's positions became increasingly aligned with officially sanctioned policy approaches, and civil society actors under its umbrella came to be 'integrally involved in (re-)producing the emerging hegemonic consensus' of turning to the market to solve the climate crisis. Taking advantage of the disputes between the fossil fuel and renewable energy factions of capital, CAN-I supports the neoliberalising project of creating a 'green economy'. While calling for an immediate end to fossil fuel subsidies, CAN-I simultaneously urges a transition to a global economy powered by renewable energy by 2050 at the latest and supports a variety of market instruments to facilitate this transition (CAN, 2015). The argument that 'green capitalism' constitutes the continuation of the neoliberalisation project (Goodman and Salleh, 2013; Kenis and Lievens, 2016) is evident in how it is being promoted not only by representatives of emerging renewable technology industries and their allies (like CAN-I), but also by institutions such as the World Bank and the IMF, as well as by other financial institutions working to commodify the natural resources that all life depends on: resources such as water (much of which has already been commodified and privatised), the atmosphere (by putting a price on carbon), forests and soil (Adelman, 2015).

Commodification entails assigning monetary value to an entity so that it can be traded in markets. While not linking its policies overtly to the neoliberal project of commodifying nature, CAN-I's support for this project is clearly evident in its 2016 Annual Report (CAN-I, 2016) and its 2015 annual policy document (CAN, 2015), where it insists on the need to accurately monitor, measure and record 'sources and sinks' of GHG emissions. Newell and Bumpus (2012) draw attention to the relationship between the seemingly 'technical exercises' of measurement and the 'intricate politics' of commodifying carbon: 'Carbon has to be rendered manageable, containable and quantifiable, fungible in value, and commensurate to be tradable as a commodity' (p. 55). CAN-I's support for this expansive neoliberalisation project that aims to marketise the entirety of nature is, perhaps, motivated by 'pragmatic' considerations regarding what CAN-I leaders perceive as realistic and achievable, but it is also actively promoted by some of the larger and well-resourced CAN-I members such as the EDF (Guldbrandsen and Andresen, 2004).

The dangers of **trasformismo** and *'passive revolution'*: questioning the CAN-I position

Radical climate justice advocates argue that CAN-I's close collaboration with official policymakers and business interests is dangerous because it can lead to the cooptation of the climate movement and can also legitimise neoliberal policies while simultaneously foreclosing more socially just solutions to climate change (Bond and Dorsey, 2010). In Gramscian terms, the *trasformismo* of CAN-I is evident in how its support of false solutions furthers current attempts by the transnational capitalist class (TCC) and its allies to rebuild and strengthen capital's hegemonic historic bloc by incorporating concerns about climate change, neutralising them and thereby achieve a 'revolution from above'. Through such a passive revolution, neoliberal capitalism can expand by commodifying nature and thereby not only 'save' itself from the current accumulation crisis (which has been ongoing since the GFC), but it can also strengthen and expand its power and reach (Goodman and Salleh, 2013; Kenis and Lievens, 2016).

Trasformismo constitutes only one of a repertoire of tactics adopted by elites to 'neutralise' antagonistic SMOs; other tactics are more coercive and include limiting non-state actor access to international conferences if they 'challenge market ideologies or neoliberal ideas', using legislation to repress 'non-violent progressive groups' and stigmatising and delegitimising progressive groups in order to limit their influence in the wider community (Smith, Plummer and Hughes 2017, pp. 4–6). Many of these tactics were used to disempower the climate movement at COP-15 in Copenhagen in 2009, where the newly formed radical climate justice alliance *CJN!* gained prominence as a major actor in the climate movement.[6]

Climate Justice Now! (CJN!)

Like the efforts to precisely date the emergence of the new environmental movement by identifying a 'big book' or a significant event, various analysts have attempted to locate a singular event marking the emergence of the radical climate justice movement. Some of the literature points to the 2007 COP-13 in Bali, when a number of NGOs broke away from CAN-I, as the origins of the radical climate justice movement's manifestation as *CJN!* (for example, refer to de Lucia, 2014; Bond, 2014; and Tokar, 2014). Analysts and commentators also identify several other significant events, such as the first known conference on the theme of 'climate justice' at the 2000 alternative people's climate summit at COP-6 in The Hague (Bond, 2014, p. 208; Tokar, 2014) and the manifesto outlining 27 *Principles of Climate Justice* developed by 14 NGOs at the 2002 United Nations World Summit on Sustainable Development in Johannesburg (Steger, Goodman and Wilson 2013). COP-15 in Copenhagen in 2009 is also widely referred to as significant in the development of the radical climate justice movement (for example, refer to Bedall and Görg, 2014; Bond, 2014, della Porta and Parks, 2014), as is the 2010 Cochabamba People's Agreement, a programme developed at an alternative climate change summit convened at the invitation of Bolivian president Evo Morales in Cochabamba after the failure of COP-15 (Angus, 2016). All these events are important in their own right as instances of climate justice activism and their contributions to the evolution of both the ideology and the strategy and tactics adopted by the radical climate justice network have been extensively analysed.[7] However, as Meyer and Rohlinger (2012) caution, the evolution of social movements is complex and occurs over a much longer period than is generally acknowledged in the literature; it should therefore come as no surprise that ideas about climate justice were being discussed widely for many years within the climate movement and the global justice movement before *CJN!* was created as an identifiable radical climate justice network. Rather than trying to identify all the significant events shaping the development of the radical climate justice movement, it would therefore be more fruitful to focus on its aims and strategies, and on the ideology informing these.

Competing ideas: radical climate justice versus climate action

The main ideas defining the radical climate justice movement have been linked to the environmental justice movement of the 1980s, a 'movement that especially emphasized the racial and class injustices of pollution in the United States' (Bond, 2014, p. 208), as well as to the GJM associated with the anti-capitalist protests against the WTO in Seattle in 1999 (Garrelts and Dietz, 2014; Steger, Goodman and Wilson, 2013; Tokar, 2014). According to Tokar (2014), the first published reference to 'climate justice' was in the 1999 report *Greenhouse Gangsters vs. Climate Justice* (TRAC, 1999), written by a group associated with the GJM. Not only does this document refer to 'climate justice', but it also draws attention to 'false solutions' and calls for a 'just transition' to a fossil fuel-free economy.

With a strong critique of corporate power, and especially of the power of the fossil fuel industry, *Greenhouse Gangsters vs. Climate Justice* identifies 'technological and market-oriented' solutions such as emission trading schemes (ETSs), as well as the Kyoto Protocol's Joint Implementation and Clean Development Mechanism (CDM), as 'false solutions', and attempts to set out 'a platform for Climate Justice' (TRAC, 1999, p. 6 and pp. 23–26). The authors of the report argue that the only way to counter the immense power of the fossil fuel corporations and their many corporate, political and institutional allies is to build a powerful grassroots movement linking social and environmental struggles around the globe. They suggest that climate justice activists join forces with allies such as: alter-globalisation activists trying to dismantle the power of corporations and global economic institutions; indigenous people and subsistence farmers in the Global South trying to defend their territories, health and livelihoods against the environmental threats of fossil fuel projects; disadvantaged groups in the Global North trying to defend their communities against toxic pollution; and workers calling for a just transition from a fossil fuel economy.[8] These suggestions are reflected in much of the thinking within the current radical climate justice movement, as well as in its operation as a network of grassroots groups working locally on specific campaigns while simultaneously demonstrating solidarity with their allies' campaigns (for example, refer to Bond, 2010; Klein, 2014; and Temper and Gilbertson, 2015).

A comparison of the principles and positions of the two extreme wings of the climate movement suggests that the two features that fundamentally distinguish the radical climate justice movement from CAN-I, and from other groups that lie in between the two extremes, are its anti-capitalism and its uncompromising ethical stance. Radical climate justice movement actors do not support measures identified as 'false solutions' by those who are affected by them; instead, they provide principled support to the least powerful and most marginalised people whose livelihoods and very lives are discounted by a neo-liberalising and totalising global capitalist economy, which also discounts the lives of other species, entire ecosystems and all future generations. In other words, the radical climate justice movement adopts an uncompromising *ethical* stance that gives it the moral high ground, and although many would argue that this is its greatest weakness, it could also be argued that this is its greatest strength if one considers the practical outcomes of 'pragmatic' approaches.

While a CAN-I member reflecting on the split between CAN and *CJN!* concluded that 'Those groups that left have become less visible in the UN process, in which CAN is now the dominant NGO player' (CAN, 2014, p. 157), an objective consideration of what CAN-I has *achieved* as a result of choosing to work cooperatively within the system to encourage reforms suggests that its position of 'dominance' in the official climate change negotiations is perhaps not that significant. COP-15 was a disaster, with many analysts agreeing that the 2009 Copenhagen Accord was a last-minute attempt to salvage a very bad situation (Carter, Clegg and Wåhlin, 2011; Parker et al., 2012). The 2015 Paris Agreement's so-called 'bottom-up approach' is just as ineffective (if not actually regressive in comparison to the Kyoto Protocol that preceded it), having yielded Intended Nationally Determined Contributions (INDCs) that lock in a global average warming of 'between 2.6 and 3.1 degrees Celsius by 2100' (Rogelj et al., 2016), 'contributions' which are, in any case, voluntary and not legally binding.[9]

It is clear that the best that global ruling elites can do given the organic crisis they face is to achieve a temporary and uneasy passive revolution; they can neither solve the 'natural limits' problem with their technical and market solutions (which, in any case, demonstrably often serve only to extract profits and not to mitigate climate change, as they claim), nor are they willing to devise and implement a real compromise with subaltern classes whose suffering intensifies as the climate crisis and all its various effects continue to unfold. While the TCC and its allies have already lost the moral battle, continuing to reveal the causes of the many injustices that exist and the worse ones that are unfolding is the radical climate justice movement's main strength. As many analysts and activists argue, it is crucial that supporters of the climate justice movement maintain constant vigilance against adopting tactics that involve making compromises that will lead to very different scenarios to those it aims to promote. Such vigilance is very difficult to maintain given the urgent nature of the challenges presented by AGW, as these put a lot of pressure on radical climate justice movement actors to make exactly these sorts of compromises. However, the stakes of ensuring that 'real solutions' that incorporate socio-economic justice and democratic concerns are adopted are very high, as the kind of society that is almost certain to result from 'false solution' reformist or neoliberal compromises will lead not only to increasing social inequalities, but to much death and destruction. One of the crucial roles that ecosocialists play within the climate movement is to consistently draw attention to what is at stake when choosing between strategies and tactics that lead to 'false' and 'real' solutions, and this is the subject of the next chapter.

Notes

1 Different analysts use a variety of criteria to disaggregate the climate movement for analytical purposes, and the terminology they invent also varies (sometimes confusingly, as different theorists sometimes use the same terms, but define them in different ways). Steger, Goodman and Wilson (2013), for instance, seem to use a criterion that hinges on the relative access to power of different groups engaged in the climate negotiation process because the two extremes they identify fall broadly into governmental and intergovernmental actors in one category and actors within civil society's climate movement in another category. They disaggregate these extremes into additional sub-categories, identifying two reform-oriented perspectives that dominate formal climate negotiations as 'market globalism' and 'elite Third-World developmentalism', both of which aim to 'protect vested interests and maximize economic growth and industrialism'. They also distinguish between three positions in the informal climate movement located within civil society: 'climate action', 'climate autonomy' and 'climate justice', providing the disclaimer that several of the groups they included in their study 'engaged with more than one of these approaches' (*ibid.*, pp. 127–128). This disclaimer constitutes another illustration of the difficulties of analytically disentangling the continuum of positions that the climate movement's component actors represent. Using climate movement participants' 'prognostic framing' (that is, what solutions they propose for addressing global warming) rather than relative power as their criterion, Wahlström, Wennerhag and Rootes (2013, pp. 102, 105) categorise climate movement actors into those who support climate justice and call for 'systemic changes involving global justice', those who call for individual 'behavioural changes' (which may be either reformist or relate to social movement 'life politics') and reformists who support policy changes within existing institutions.

2 The importance of this particular line of division is acknowledged by Garrelts and Dietz (2014, p. 7), who point out that there are serious disagreements between groups and actors who are supportive (and sometimes actively engaged in facilitating) 'the concrete implementation of existing climate policy instruments' and groups and actors who reject all such compromises. Similar analytical distinctions are also made by other analysts, some of whom discuss the environmental and climate movements by referring explicitly to Robert Cox's important distinction between problem-solving and critical approaches as well as to the Gramscian notion of hegemony (for example, refer to Elliott, 2004, pp. 224–226 and de Lucia, 2014).

3 Ideological positions also influence tactical actions, as was evident the day after the 2014 People's Climate March, when the radical wing of the climate justice movement organised and participated in a much smaller but more militant 'breakaway Flood Wall Street action', with the aim of drawing attention to the links between 'Wall Street' (representing capitalism) and the climate crisis (Giacomini and Turner, 2015).

4 Some climate activists are much more forthcoming than others about what sort of post-capitalist future they envisage as possible or desirable, which is another instance of the 'messiness' of real life that defies containing either ideas or people into the convenient, neat categories that would expedite analysis; for example, widely respected climate activist Naomi Klein has been critiqued on this issue by some climate justice movement actors but others, with a more nuanced understanding of strategy and tactics, defend her against these critiques (for example, refer to Smith and Foster, 2017). This debate performs a vital function (as do all such debates within the climate movement, as long as they are conducted in ways that are respectful) by contributing to a deeper understanding of the issues by all parties concerned and by others following it and thinking about the issues raised.

5 As Connelly (2007) points out, 'sustainable development' is a contested concept, and different definitions of 'sustainability' and 'sustainable development' favour a variety of political projects (see also Brown, 2016). Definitions of sustainable development that locate it within a context of 'green capitalism' are sometimes used to support neoliberal governmentality while other definitions support radical socialist egalitarianism and yet others support a range of positions in between.

6 Much has been written about the very eventful 2009 climate change negotiations at COP-15 in Copenhagen, riddled as they were with various intrigues and subterfuges that led to the ultimate failure of the climate change conference and resulted in a very weak and ineffectual final document, the 'Copenhagen Accord'. Government representatives who had been excluded from the secret negotiations refused to adopt this document; instead, the Copenhagen Accord was merely 'noted'. Refer to Carter, Clegg and Wåhlin (2011), Mason and Askins (2013) and Wahlström, Wennerhag and Rootes (2013) for accounts about different aspects relevant to COP-15 events and outcomes.

7 The *Routledge Handbook of the Climate Change Movement* (Dietz and Garrelts, 2014) is an excellent introductory resource, and Steger, Goodman and Wilson (2013) also provide a detailed account of the climate movement's components and some of the key events in its development.

8 A more recent publication by this group, *A–Z of Green Capitalism* (Corporate Watch, 2016), is available on the Corporate Watch website and provides explanations of the meaning and significance of complex concepts relevant to climate change and climate justice in simple language.

9 The 'bottom-up' approach *sounds* democratic but is, in true Orwellian fashion, quite the opposite since it was adopted at COP-21 against the wishes of many other governments and only after many years of US insistence that legally binding GHG emission targets be replaced by voluntary and non-binding INDCs.

References

Adams, M. (2017). Extreme Weather Just Might Encourage Us to Get Our Act Together On Global Warming. The Conversation, 9 May. Available at https://theconversation.com/extreme-weather-just-might-encourage-us-to-get-our-act-together-on-global-warming-76679.

Adelman, S. (2015). Tropical Forests and Climate Change: A Critique of Green Governmentality. International Journal of Law in Context, 11(2), 195–212.

Agrawala, S. (1999). Early Science-Policy Interactions in Climate Change: Lessons from the Advisory Group on Greenhouse Gases. Global Environmental Change, 9(2), 157–169.

Angus, I. (2016). Facing the Anthropocene: Fossil Capitalism and the Crisis of the Earth System. New York: Monthly Review Press.

Bedall, P. and Görg, C. (2014). Antagonistic Standpoints: The Climate Justice Coalition Viewed in Light of a Theory of Societal Relationships with Nature. In Dietz, M. and Garrelts, H. (eds), Routledge Handbook of the Climate Change Movement. New York: Routledge, 44–65.

Betsill, M. (2002). Environmental NGOs Meet the Sovereign State: The Kyoto Protocol Negotiations on Global Climate Change. Colorado Journal of International Law & Policy, 13(1), 49–64.

Blühdorn, K. (2017). Post-Capitalism, Post-Growth, Post-Consumerism? Eco-Political Hopes Beyond Sustainability. Global Discourse, 7(1), 42–61.

Bond, P. (2010). Maintaining Momentum after Copenhagen's Collapse: Seal the Deal or 'Seattle' the Deal? Capitalism Nature Socialism, 21(1), 14–27.

Bond, P. (2014). Climate Justice In, By, and For Africa. In Dietz, M. and Garrelts, H. (eds), Routledge Handbook of the Climate Change Movement. New York: Routledge, 205–221.

Bond, P. and Dorsey, M.K. (2010). Anatomies of Environmental Knowledge and Resistance: Diverse Climate Justice Movements and Waning Eco-Neoliberalism. Journal of Australian Political Economy, 66, 286–316.

Brown, T. (2016). Sustainability as Empty Signifier: Its Rise, Fall, and Radical Potential. Antipode, 48(1), 115–133.

Burgmann, V. and Baer, H.A. (2012). Climate Politics and the Climate Movement in Australia. Melbourne, VIC: Melbourne University Press.

Carter C., Clegg, S. and Wåhlin, N. (2011). When Science Meets Strategic Realpolitik: The Case of the Copenhagen Climate Change Summit. Critical Perspectives on Accounting, 22, 682–697.

Climate Action Network (CAN) (2014). Small Window, Bright Light: A History of Climate Action Network. Available at www.climatenetwork.org/sites/default/files/can_history_booklet_web.pdf.

Climate Action Network (CAN) (2015). The Paris Package: A Springboard for Sustained, Transformative Change. CAN Annual Policy Document 2015. Available at www.climatenetwork.org/sites/default/files/can_annual_policy_document_the_paris_package_november_2015.pdf.

Climate Action Network International (CAN-I) (2016). Climate Action Network 2015 Annual Report. Available at www.can-network.org/files/CAN%20annual%20report%202015.pdf.

Climate Action Network International (CAN-I) (n.d.). Climate Action Network International homepage. Available at http://climatenetwork.org/.

Conca, K. (2004). Ecology in an Age of Empire: A Reply to (and Extension of) Dalby's Imperial Thesis. *Global Environmental Politics*, 4(2), 12–19.

Connelly, S. (2007) Mapping Sustainable Development as a Contested Concept. *Local Environment*, 12(3), 259–278.

Corporate Watch (2016). *A-Z of Green Capitalism*. Corporate Watch. Available at https://corporatewatch.org/publications/2016/A-to-Z-of-green-capitalism.

Cramer, K.J. (2017). The Great American Fallout: How Small Towns Came to Resent Cities. *The Guardian*, 19 June. Available at www.theguardian.com/cities/2017/jun/19/americas-great-fallout-rural-areas-resent-cities-republican-democrat.

Dalby, S. (2004). Ecological Politics, Violence, and the Theme of Empire. *Global Environmental Politics*, 4(2), 1–11.

della Porta, D. and Parks, L. (2014). Framing Processes in the Climate Movement: From Climate Change to Climate Justice. In Dietz, M. and Garrelts, H. (eds), *Routledge Handbook of the Climate Change Movement*. New York: Routledge, 19–30.

de Lucia, V. (2014). The Climate Justice Movement and the Hegemonic Discourse of Technology. In Dietz, M. and Garrelts, H. (eds), *Routledge Handbook of the Climate Change Movement*. New York: Routledge, 66–83.

Dietz, M. and Garrelts, H. (eds) (2014). *Routledge Handbook of the Climate Change Movement*. New York: Routledge.

Doherty, K.L. and Webler, T.N. (2016). Social Norms and Efficacy Beliefs Drive the Alarmed Segment's Public-Sphere Climate Actions. *Nature Climate Change*, 6(9), 879–884.

Elliott, L. (2004). *The Global Politics of the Environment* (2nd edn). New York: New York University Press.

Garrelts, H. (2014). Organization Profile – Climate Action Network International. In Dietz, M. and Garrelts, H. (eds), *Routledge Handbook of the Climate Change Movement*. New York: Routledge, 237–239.

Garrelts, H. and Dietz, M. (2014). Introduction: Contours of the Transnational Climate Movement – Conception and Contents of the Handbook. In Dietz, M. and Garrelts, H. (eds), *Routledge Handbook of the Climate Change Movement*. New York: Routledge, 1–16.

Giacomini, T. and Turner, T. (2015). The 2014 People's Climate March and Flood Wall Street Civil Disobedience: Making the Transition to a Post-Fossil Capitalist, Commoning Civilization. *Capitalism Nature Socialism*, 26(2), 1–19.

Goodman, J. and Salleh, A. (2013). The 'Green Economy': Class Hegemony and Counter-Hegemony. *Globalizations*, 10(3), 411–424.

Guldbrandsen, L.H. and Andresen, S. (2004). NGO Influence in the Implementation of the Kyoto Protocol: Compliance, Flexibility Mechanisms, and Sinks. *Global Environmental Politics*, 4(4), 54–75.

Gupta, J. (2010). A History of International Climate Change Policy. *WIREs Climate Change*, 1, 636–653.

Hausknost, D. (2017). There Never Was a Categorical Ecological Imperative: A Response to Ingolfur Blühdorn. *Global Discourse*, 7(1), 62–65.

Hoffarth, M.R. and Hodson, G. (2016). Green on the Outside, Red on the Inside: Perceived Environmentalist Threat as a Factor Explaining Political Polarization of Climate Change. *Journal of Environmental Psychology*, 45, 40–49.

Hornsey, M.J., Harris, E.A., Bain, P.G. and Fielding, K.S. (2016). Meta-Analyses of the Determinants and Outcomes of Belief in Climate Change. *Nature Climate Change*, 6, 622–626.

Kenis, A. and Lievens M. (2016). Greening the Economy or Economizing the Green Project? When Environmental Concerns are Turned into a Means to Save the Market. *Review of Radical Political Economics*, 48(2), 217–234.

Klein, N. (2014). *This Changes Everything: Capitalism vs the Planet*. London: Allen Lane.

Kraft, M.E. (2000). U.S. Environmental Policy and Politics: From the 1960s to the 1990s. *Journal of Policy History*, 12(1), 17–42.

Land is Life, Indigenous Environmental Network, Society for Threatened Peoples, Global Forest Coalition, Pacific Indigenous Peoples Environment Coalition, Carbon Trade Watch, CORE, Global Justice Ecology Project World Rainforest Movement, CEALP, ICTP and Transnational Institute. (n.d.). *Indigenous People's Guide: False Solutions to Climate Change*. Available at earthpeoples.org.

Lukacs, M. (2017). Neoliberalism has Conned Us into Fighting Climate Change as Individuals. *The Guardian*, 17 July. Available at www.theguardian.com/environment/true-north/2017/jul/17/neoliberalism-has-conned-us-into-fighting-climate-change-as-individuals.

Mason, K. and Askins, K. (2013). COP15 and Beyond: Politics, Protest and Climate Justice. *ACME: An international E-Journal for Critical Geographies*, 12(1), 9–22.

McCright, A.M., Charters, M., Dentzman, K. and Dietz, T. (2016). Examining the Effectiveness of Climate Change Frames in the Face of a Climate Change Denial Counter-Frame. *Topics in Cognitive Science*, 8, 76–97.

McKibben, B. (2012). Connecting the Dots of this Climate Change Crisis. *The Guardian*, 4 May. Available at www.theguardian.com/commentisfree/cifamerica/2012/may/04/connecting-dots-climate-change-crisis.

Meyer, D.S. and Rohlinger, D.A. (2012). Big Books and Social Movements: A Myth of Ideas and Social Change. *Social Problems*, 59(1), 136–153.

Newell, P. and Bumpus, A. (2012). The Global Political Ecology of the Clean Development Mechanism. *Global Environmental Politics*, 12(4), 49–67.

Parker, C.F., Karlsson, C., Hjerpe, M. and Linnér, B-O. (2012). Fragmented Climate Change Leadership: Making Sense of the Ambiguous Outcome of COP-15. *Environmental Politics*, 21(2), 268–286.

Robertson, T. (2008). 'This is the American Earth': American Empire, the Cold War, and American Environmentalism. *Diplomatic History*, 32(4), 561–584.

Rogelj, J., den Elzen, M., Höhne, N., Fransen, T., Fekete, H., Winkler, H., Schaeffer, R., Sha, F., Riahi, K. and Meinshausen, M. (2016). Paris Agreement Climate Proposals Need a Boost to Keep Warming Well Below 2°C. *Nature*, 534, 631–639.

Sale, K. (1993). *The Green Revolution: The American Environmental Movement, 1962–1992*. New York: Hill and Wang.

Smith, J., Plummer, S. and Hughes, M.M. (2017). Transnational Social Movements and Changing Organizational Fields in the Late Twentieth and Early Twenty-First Centuries. *Global Networks*, 17(1), 3–22.

Smith, R. and Foster, J.B. (2017). The Significance of Naomi Klein: An Ecosocialist Exchange. Climate & Capitalism, 4 May. Available at http://climateandcapitalism.com/2017/05/04/significance-naomi-klein-ecosocialist-exchange.

Spence, A., Poortinga, W. and Pidgeon, N. (2012). The Psychological Distance of Climate Change. *Risk Analysis*, 32(6), 957–972.

Steffen, W., Hughes, L., Alexander, D. and Rice, M. (2017). *Cranking Up the Intensity: Climate Change and Extreme Weather Events*. Climate Council of Australia. Available at www.climatecouncil.org.au/cranking-intensity-report.

Steger, M.B., Goodman, J. and Wilson, E.K. (2013). *Justice Globalism: Ideology, Crises, Policy*. London: Sage Publications.

Tansey, R. (2013). *The COP19 Guide to Corporate Lobbying: Climate Crooks and the Polish Government's Partner in Crime*. Corporate Europe Observatory & Transnational Institute. Available at http://corporateeurope.org/blog/cop19-guide-corporate-lobbying.

Temper, L. and Gilbertson, T. (eds) (2015). *Refocusing Resistance to Climate Justice: COPing In, COPing Out and Beyond Paris*. EJOLT Report, 23. Available at www.ejolt.org/2015/09/refocusing-resistance-climate-justice-coping-coping-beyond-paris/.

Tanuro, D. (2013). *Green Capitalism: Why It Can't Work*. London: The Merlin Press Ltd.

Tokar, B. (2014). *Toward Climate Justice: Perspectives on the Climate Crisis and Social Change*. Porsgrunn, Norway: New Compass Press.

Transnational Resource and Action Center (TRAC) (1999). Greenhouse Gangsters vs. Climate Justice. CorpWatch, 1 November. Available at http://corpwatch.org/article.php?id=1048.

Wahlström, M., Wennerhag, M. and Rootes, C. (2013). Framing 'The Climate Issue': Patterns of Participation and Prognostic Frames among Climate Summit Protesters. *Global Environmental Politics*, 13(4), 101–122.

World Meteorological Organization (WMO) (2016). (Un)Natural Disasters: Communicating Linkages between Extreme Events and Climate Change. *World Meteorological Organization Bulletin*, 65(20).

6   Competing ideas

Ecosocialist strategy and tactics in the struggle for climate justice

Introduction

The overview of the key points of contention in academic theoretical debates between ecosocialists in Chapter 3 manifests at the level of activism on questions regarding strategy and tactics. The discussion of debates about strategy and tactics presented in this chapter focuses specifically on how these relate to ecosocialist contributions to furthering the aims of the radical climate justice movement. After a general overview of the kinds of issues generating discussion amongst ecosocialists, the ways in which debates about strategy and tactics unfold are illustrated with reference to three case studies. The first case study involves a debate about whether or not ecosocialists should support carbon trading as a means of reducing GHG emissions; the second case study discusses the reasoning underlying the decision of ecosocialist activists to support a strike by workers in the fossil fuel industry; the third case study examines the debate between ecosocialists and climate justice activists in the wider climate justice movement on appropriate strategies and tactics to be adopted at the 2015 COP-21 climate negotiations in Paris.

These debates were selected as foci because the first demonstrates ecosocialist views on market mechanisms (the 'false solutions'/'real solutions' issue within the climate justice movement discussed in Chapter 5), the second demonstrates ecosocialist views about the role of the working class in struggles to mitigate and adapt to climate change and the third demonstrates ecosocialist attitudes towards the formal climate change institutions and the extent to which their ideas influence the broader climate movement. Using elements of the MHS Redux to guide the analysis, the discussion in this chapter begins with a brief overview of the role of ecosocialists within the climate justice wing of the climate movement.

Social dynamics: ecosocialism as a distinct current within the climate justice movement

One indicator that ecosocialist ideas have gained ground within the climate movement is the way in which they have shifted from debates within the academy (as discussed in Chapter 3) and are now used by many climate movement activists to inform their analyses of the causes of, and potential solutions to, climate change. The adoption of ecosocialist ideas is also evident in the fact that various explicitly named 'ecosocialist coalitions' are currently active in both North America and Europe, with the North American groups only emerging in their present coherent form after the first decade of the twenty-first century. Two events point to the year 2007 as being significant in this respect in that it signalled efforts by activists to form coalitions calling for 'system change' (rather than system reform): the formation of the *Climate Justice Now!* (*CJN!*) coalition in December 2007 at COP-13 in Bali (as discussed in Chapter 5) and, predating the formation of this coalition by two months, the attempt by a small group of ecosocialists to establish an international ecosocialist coalition.

Attempts to form an international ecosocialist coalition date back to October 2007, when more than 60 ecological activists from 12 countries met in Paris and inaugurated the Ecosocialist International Network (EIN) (Climate & Capitalism, 2007; EIN, 2007). Ian Angus, one of the founders of this network, acknowledges that the EIN had 'obvious weaknesses' – in particular, its lack of representativeness: 'the great majority of attendees were from left groups in western Europe, and only a handful came from the global south' (Angus, 2008). In a 2016 interview, Ian Angus also emphasised his conviction that the EIN's greatest weakness was that it was a 'top-down' attempt to form an organisation, and that for this reason it was bound to fail (personal communication, 2016). The EIN nevertheless constituted what Angus (2008) describes as 'a big step forward for the ecosocialist current' in that it signalled an initial attempt to build a movement that could act on the emerging 'ecosocialist project' outlined in the *First Ecosocialist Manifesto* authored by Michael Löwy and Joel Kovel and published in 2001.[1] One outcome of the 2007 EIN meeting was the *Draft Second Ecosocialist Manifesto* developed by Kovel, Löwy and Angus, and publicly discussed on a 'yahoogroups' EIN forum convened by a four-person committee whose members included women, people from the Global South (Brazil) and younger people (Angus, 2008). The document that emerged from these discussions, the Belem Ecosocialist Declaration, was endorsed 'by hundreds of people from dozens of countries' (Löwy, 2015, p. xiii) and presented at the 2009 World Social Forum (WSF) in Belem, Brazil (EIN, 2009).[2]

While the EIN represents a significant milestone in the development of ecosocialist activism, as discussed in Chapter 5, there is no simple 'origin story' through which one can trace the history of social movements or groups such as the System Change Not Climate Change (SCNCC) ecosocialist coalition. The complex origins and alliances of the contemporary 'ecosocialist current' within the broader climate movement are evident in how, according to information in an SCNCC pamphlet, some ecosocialists were involved in the alter-globalisation movement that emerged in the 1990s as well as in how ecosocialist ideas and solidarity actions are informed by indigenous environmental justice activist groups and by grassroots movements such as the Landless Workers' Movement in Brazil and the food sovereignty movement repres- ented by La Via Campesina, a worldwide organisation uniting peasant and subsistence farmers (SCNCC, n.d.). While all this history constitutes the background to the formation of SCNCC, the coalition itself emerged in its present form after the 20 April 2013 Ecosocialist Conference organised by the 'Ecosocialist Contingent' that had participated in the 17 February 2013 Forward on Climate March campaign, a mass demonstration held in Washington, DC against the Keystone XL tar sand pipeline (Angus et al., 2013; Climate & Capitalism, 2013).

SCNCC describes itself as: 'a joint Canadian and US coalition of ecoso- cialists and fellow travellers united in the belief that capitalism is driving climate change and that a radical international grassroots movement can stop it' (SCNCC, 2017b). Having participated in various campaigns as the 'Ecoso- cialist Contingent' and then 'System Change Not Climate Change' for 'roughly a year', ecosocialists organised a conference in New York City in 2013 that featured speakers 'from across the leftish spectrum', some of whom went on to form SCNCC as it currently exists (SCNCC, n.d.; see also Ecosocialist Conference, 2013, Ecosocialist Contingent, 2013; and Zill and Ware, 2014).[3] The SCNCC coalition is organised in chapters located in different cities or regions in the US and Canada, with members participating in monthly conference calls to coordinate their activities and discuss future plans (SCNCC, n.d.). While some members of SCNCC belong to other political organisations, such as the International Socialist Organization, Solid- arity and the Green Party, others are independent and, in recognition of the differences between members, the coalition focuses on 'areas of agreement' in their struggles for environmental and social justice (*ibid.*, p. 7). The issues on which SCNCC members agree are embodied in six 'points of unity' that were developed 'as a step towards overcoming the sectarianism that has dogged the Left' (*ibid.*).

The points of unity that SCNCC members agree on are that they: see capitalism as the cause of the current ecological crisis; see exploitation in all its forms as being intricately linked, and therefore oppose 'all forms of oppression including racism, sexism, xenophobia, homophobia and trans- phobia' in their struggles to stop 'the exploitation and destruction of the planet'; share a vision of a future 'free, just, and equitable [society] that fosters human creativity and productivity while healing the rifts generated by capit- alism among people and between human society and the earth's ecology'; are open to supporting reformist campaigns such as those initiated by organisa- tions such as 350.org, Friends of the Earth (FoE) and Greenpeace in order to address 'immediate ecological concerns' while working towards the funda- mental system change that is required for building a truly sustainable society in the long term; oppose 'green capitalism' and support alternative political formations such as grassroots movements rather than 'the capitalist-controlled two-party system'; and adopt a non-sectarian stance 'in order to build as strong an environmental movement as possible' (*ibid.*, pp. 7–8). In Gramscian terms, these points of unity represent ecosocialist attempts to work with other groups and individuals in constructing a counter-hegemonic bloc to challenge capitalism. The discussion in this chapter analyses these contributions to the global ecosocialist project of conducting a Gramscian war of position on the terrain of civil society, whereby dominant ideas regarding the causes of, and solutions to, the current organic crisis are challenged.

Material capabilities: ecosocialist resources

While social movements tend to have very limited access to material resources in class societies, in the past few decades they have had access to one important resource: online media and communication tools. Ecosocialist ideas are developed, refined and disseminated through various online media, including websites, video conferences, social media such as Facebook and Twitter and video recordings of meetings, conferences and book launches that are shared via online media platforms such as YouTube and Vimeo. The use of ICTs can therefore be seen as one tactic whereby ecosocialists engage in the war of position to present their system-critical arguments about the causes of, and solutions to, the current organic crisis of global capitalism. The two most significant ecosocialist websites in North America are Climate & Capitalism (climateandcapitalism.com) and SCNCC's website (systemchangenotclimatechange.org).

Examples of the sort of material published on the aforementioned websites include articles about climate science and about environmental campaigns as well as articles outlining ecosocialist theory and action, including crucial debates about strategy and tactics. While the SCNCC website largely acts as a clearing house for climate change- and climate justice-related information published elsewhere (with occasional original articles not published elsewhere), Climate & Capitalism publishes many original articles written by ecosocialist activists and academics. Collaboration between various North American ecosocialist groups and projects is evident in the fact that the SCNCC website has direct links to articles published on the Climate & Capitalism website and in Climate & Capitalism's alliance with the prominent ecosocialist journal *Monthly Review*, with which it explores 'ways to increase both technical and political collaboration' and shares technical resources (Angus, 2012). There are also links on both the SCNCC and Climate & Capitalism websites to articles published on a variety of other alternative political organisation websites, such as those of Socialist Project (socialistproject.ca) and Solidarity (solidarity-us.org). While the Climate & Capitalism website offers readers opportunities to comment on the content it publishes and the SCNCC website does not, the SCNCC added an online forum to its website in June 2017 to facilitate discussion and the exchange of ideas. It is through such online communications that ecosocialist strategy and tactics are actively debated and co-developed in a way that is available to a global audience and invites input from this audience.

Social dynamics in building counter-hegemony: ecosocialist strategies and tactics

In direct opposition to the moderate climate action movement's generally reformist positions, the radical climate justice movement's understanding of capitalism as the cause of the current ecological, social, economic and political crises leads it to conclude that climate change and other equally important problems can only be addressed by fundamental 'system change'. Thus the radical climate justice movement sees the solutions to the issue of climate change proposed by supporters of the moderate climate action movement as being fundamentally 'false solutions', although some of these solutions are perceived as necessary interim or transitionary measures that should be pursued to mitigate further climate change. As some critical theorists emphasise, however, this support of transitionary measures is risky as it opens up avenues for the moderate climate movement to coopt the radical climate justice movement and can ultimately lead to *trasformismo* (for example, see de Lucia, 2014). One of the central challenges faced by climate justice activists is, therefore, how to build alliances with other climate movement actors while avoiding cooptation and continuing to focus their efforts on promoting 'real solutions' that are not only effective in maintaining a habitable planet but are also socially just. Ecosocialist analyses and theoretical contributions are very valuable in this respect. One of the articles written by prominent ecosocialist activist-academic and author Chris Williams addresses the issue of strategy and tactics and refers to an example of the sort of tactics that build a stronger and more effective climate justice movement as opposed to tactics that legitimise the existing system of unequal power relationships. Williams's discussion is summarised below as an introduction to the issue of ecosocialist ideas about strategy and tactics.

Ecosocialist activism: strategy and tactics

In a 2013 article entitled 'Strategy and Tactics in the Environmental Movement', published on the Climate & Capitalism website, Chris Williams provides the following definitions from the *New Webster's Comprehensive Dictionary*.

> *Tactics*: the science and art of using a fighting force to the best advantage having regard to the immediate situation of combat. *Strategy*: the science and art of conducting a military campaign in its large-scale and long-term aspects.
>
> (Williams, 2013)

Referring to a debate sparked by 'prominent climate blogger Joseph Romm', who advised that Naomi Klein's soon-to-be-published book, *This Changes Everything: Capitalism vs. the Climate*, be ignored on the grounds that Klein's views are 'filled with contrarian "media bait" statements devoid of substance',

Williams argues that the broader debates boil down to this 'single and vitally important question' (*ibid.*):

> what is the most effective terrain, and with which combination of troops and allies, should the environmental movement engage with opposing forces in order to emerge victorious?

Rephrasing this question later on in the article, Williams asks (*ibid.*):

> How can we both fight for meaningful change right now (tactics) that simultaneously helps build the movement and brings us closer to our larger, more long-term goals (strategy)? How do we differentiate between effective tactics that supplement our overall strategy, versus those that lead us up blind alleys?

Acknowledging the shortcomings of the analyses of groups such as 350.org regarding the causes of climate change, and the limited potential of reformist efforts such as fossil fuel divestment campaigns to effect the societal shifts necessary to decisively address climate change, Williams argues that the most important contribution socialists and radicals can make is nevertheless to join such struggles and help to play a role in how the many new organisations popping up around the issue of climate change develop.

> [T]he most important thing is to dive into the resistance as and where it currently exists and consistently engage with the fight for … immediate goals of campaigns such as those calling for universities and pension funds to divest from fossil fuels and attempts to shut down the KXL oil pipeline.
>
> (Williams, 2013)

At the same time, however, Williams warns that radical activists should hold 'no illusions' that participating in such campaigns is sufficient. While it is important to achieve short-term and limited victories because of the optimism and motivation these engender, he emphasises that *how* these victories are achieved (in other words, which tactics are followed) is crucial, and that ecosocialists should be participating in discussions about strategy and tactics. Williams illustrates his article with reference to different types of legal contestations:

> A court victory achieved by NGO lawyers working in a social vacuum is completely different to a court victory achieved on the backs of mass mobilization, as illustrated by the civil rights movement … We should be part of all the discussions now going on in the movement about tactics and strategy, suggest alternatives, make the case for actions that will draw in more participants, and create links with frontline communities of color

and indigenous rights, while working with the bigger organizations where we are able. Where we have criticisms, we should voice them; in my experience they will likely find a strong echo. The ever more desperate ecological and economic situation is in itself driving people toward the need for more radical, systemic change.

(Williams, 2013)

As these extracts demonstrate, ecosocialists try to further their overall strategic aim of helping to bring about radical system change by participating in public debates such as this one. Williams' contribution in this article demonstrates that ecosocialists understand one of their primary roles within the wider climate justice and global justice movements as being to support and join with existing campaigns that attempt to mobilise more people and build a large and effective climate movement while simultaneously engaging in discussions with their fellow activists about the causes of, and solutions to, climate change in ways that provoke critical thinking about these issues.

The tactic of joining existing organisations and groups that are working on a variety of social justice issues (rather than creating new, specifically ecosocialist, organisations) is also consistently emphasised by other ecosocialists such as Ian Angus (2011; personal communication, 2016), Benjamin Silverman (Angus et al., 2013) and Gemma Weedall (2013). Ian Angus repeats this message consistently in many of his presentations and also emphasised it in interviews with the author, during which he explained that there is no 'ecosocialist movement' as such but that ecosocialists are generally members of groups engaged in a variety of social justice and ecological struggles and work within wide coalitions. As Ian Angus stated in his keynote presentation at the Climate Change Social Change Conference held in Melbourne in 2011:

[one] lesson we can learn from the 20th century is that monolithic socialist grouplets [sic] do not turn into mass movements. They stagnate and decay, they argue and they split, but they don't change the world. So I want to emphasise that I am not urging you to rush out and found yet another sect. Ecosocialism is not a separate organisation, it is a movement to win existing red and green groups and individuals to an ecosocialist perspective.

(Angus, 2011, p. 14)

The seriousness with which ecosocialists approach the challenge of formulating appropriate strategies and tactics is evidenced in the many other ongoing discussions related to these issues. In 2014, for example, the group Solidarity established an 'Ecosocialist Working Group' that formulated six questions (later adding a seventh) around which to focus such discussions (Solidarity, 2015; see also SCNCC, 2017a, where a link is provided to the page introducing these questions on the Solidarity website). As shown in Table 6.1 (which presents summary versions of the original questions organised into strategic

Table 6.1 Summary of ecosocialist questions regarding strategy and tactics

---

Questions about strategy
'How does ecosocialist politics differ from traditional socialist and labor politics?'
'What role do science, technology, labor productivity and production play in the transition from capitalism to ecosocialism, also in an ecosocialist society after the transition?'

Questions about tactics
'How does the ecological crisis affect the orientation of [workers'] unions and their place in the class struggle?'
'How, if scaling back production is necessary, will ecosocialist strategy remain committed to meeting human needs?'
'What ideas do ecosocialists raise in the climate change movement? Are James Hansen's proposals (for example, advocacy of a "carbon tax" rather than "cap and trade") in some form useful for ecosocialist transitional demands, or are they simply an attempt to solve the ecological crisis within the context of capitalism?'
'What kinds of cooperatives that can be built today might be able to teach us something about a post-capitalist world? What role, if any, should ecosocialists seek to play in these communities?'
'Is it possible for left governments in developing countries to pursue a more egalitarian social project, in the context of a global economy that continues to be dependent on extractivism, without violating basic ecological principles – in particular the demands of their own indigenous populations?'

Source: Adapted from material published on Solidarity website (2015).

and tactical categories), two of the questions are primarily related to strategy because they interrogate what ecosocialist long-term goals should be, while the rest are more concerned with tactical issues.[4] Responses to these questions by members of Solidarity's Ecosocialist Working Group and by two other authors were also published on its website (refer to Becker, 2014; Bloom 2015; Engel-DeMauro, 2015; and Feeley, 2015), and all the respondents make interesting points. However, it is the questions themselves that are discussed in more detail here as they reveal the kinds of issues that are contested amongst ecosocialists.

Ecosocialist strategy: building a powerful climate movement

Question 5 in Table 6.1, which relates to what ideas ecosocialists should be raising in the climate movement, emerges in a number of guises in the debates published on Climate & Capitalism, including in debates that emerged after the publication of Naomi Klein's 2014 book *This Changes Everything: Capitalism vs the Climate*. John Bellamy Foster and Brett Clark (2015) address both liberal and radical critiques of Klein's book about climate change, pointing out that liberal critics aim to 'rein in her arguments' and either impose interpretations that 'refashion her message' so that it does not question the permanency of capitalism or, failing that, they attempt to discredit her analysis entirely on the grounds that it is 'simplistic' or 'idealistic' (with one critic even referring to Klein as an 'idiot'). Addressing socialist critiques that Klein's argument is liberal rather than radical and that it has little to say about the working class and does not go far enough in its critique of capitalism (Smith, 2014; Smith and Foster, 2017), Foster and Clark (2015) argue that Klein's ambivalent stance, which fails to go as far as calling for socialism, is strategic.

> Her aim at present is clearly confined to the urgent and strategic – if more limited – one of making the broad case for System Change Not Climate Change. Millions of people, she believes, are crossing or are on the brink of crossing the river of fire. Capitalism, they charge, is now obsolete, since it is no longer compatible either with our survival as a species or our welfare as individual human beings … It is this burgeoning global movement that is now demanding anti-capitalist and post-capitalist solutions. Klein sees herself merely as the people's megaphone in this respect.
>
> (Foster and Clark, 2015)[5]

Similar acknowledgements of the importance of Klein's contribution to the climate justice movement, and the wider global justice movement, are made by other ecosocialists (for example, Foran, 2014; Hornick, 2015, 2017; Riddell, 2014). Brad Hornick's discussion of the importance of Klein's work emphasises its revolutionary nature with reference to Marx's critique of philosophy.

> Marx's *Theses on Feuerbach* is a critique of both idealism and mechanical materialism. The theme that reconciles the two – the admonition to make history while studying it – is a crucial aspect of the revolutionary formula. The notion that Klein's book emerges out of personal and political engagement is a confirmation that 'radical' analysis includes the type of embeddedness in praxis that Marx so brilliantly defined in this piece as well as in the *Economic and Philosophical Manuscripts* of 1844.
>
> (Hornick, 2015)

The positive responses to Klein's work, which address and deflect potentially divisive criticisms within the climate justice activist community, indicate that ecosocialists are serious about participating in the wider project of building a strong counter-hegemonic movement and also that they understand that Klein's authority within the 'Blockadia' (protest) movement derives from her dual role as activist-theorist and respect her for this.

Unlike some of her socialist critics, Klein is an organic intellectual in that she is embedded and emotionally involved in the struggles she writes about, and Brad Hornick suggests that this is what all ecosocialists should be doing.

> What I would like to emphasize in this review, is that Klein's mode of writing as embodied experience within the crisis itself grants it considerable power, and at the same time marks a departure from a problem

central to both the mainstream natural and social sciences and to the relationship between the two. This is a difference between impersonal,
objective observation and a kind of praxis – the street-level fighter's
interweaving of theory and practice. While there are many academic
writers who have paid their dues and produced much more exacting
accounts of the connections between ecology and capitalism, few can say
that they have contributed so directly to a relatively consequential
'movement,' (amidst the miserable dearth of North American radical
organizing), whilst doing so.

(Hornick, 2015)

Hornick's argument aligns with Chris Williams' (2013) advice that ecosocialists 'dive into the resistance as and where it exists', and also demonstrates a
determined effort to shift ecosocialist debates in directions that are less sectarian. However, the matter of 'principled opposition' emerges more forcefully when it comes to other tactical questions, such as the debate about
whether or not ecosocialists should support market mechanisms such as
carbon trading and carbon taxes.

Case study 1: ecosocialist tactics – are market mechanisms acceptable transitional demands?

One of the reasons that ecosocialists engage in debates about whether or not
to support climate movement actor calls for market measures such as carbon
trading and carbon taxes as a means of reducing GHG emissions (as represented by Question 5 in Table 6.1) is because these are perceived as 'false
solutions' that may lead to *trasformismo*. Avoiding *trasformismo* involves being
alert to what de Lucia (2014, p. 67) refers to as the 'boundary line' between
'system-critical' and 'system-friendly' interpretations and approaches to key
contested concepts within climate politics, and can also be applied to strategic
and tactical choices. The debate about carbon trading between prominent
climate justice advocate Robin Hahnel and ecosocialist Nicholas Davenport
demonstrates that activists working within ecosocialist organisations are
clearly aware of the dangers of *trasformismo*, as well as of the need to take
actions that further their larger goal of working towards the establishment of
ecosocialism rather than supporting actions that logically imply further
entrenching capitalist relations of production.

The debate analysed here originated with an article written by Hahnel and
published on 4 November 2013 in the journal *New Politics*. In this article,
entitled 'An Open Letter to the Climate Justice Movement', Hahnel (2013)
advocates that the climate justice movement support international carbon
cap-and-trade market mechanisms, a measure that goes against climate justice
movement assessments that market-based climate change solutions are 'false
solutions' (Dietz, 2014). Hahnel's stated reasons for adopting this position
despite being 'a long-time advocate of both climate justice and fundamental

system change' include the urgency of what the science is telling us about the need to reduce anthropogenic GHG emissions 'dramatically over the next decade' and the reality of what is politically feasible within the context of a global capitalist system which, Hahnel contends, is not likely to be seriously challenged in the immediate future. Patrick Bond (2013) points out that this is not the first time Hahnel raises this issue, having previously published at least two articles (Hahnel, 2012a, 2012b) that sparked similar debates within the climate justice movement. Hahnel's 2013 open letter again provoked responses and debates amongst climate justice movement actors, and Davenport's (2014) response is discussed in detail as it not only refutes Hahnel's argument at a theoretical and empirical level, but it also demonstrates the sort of nuanced analysis that informs ecosocialist understandings of system-critical tactics that are essential for avoiding *trasformismo*.[6]

While expressing his respect for Hahnel as a fellow climate justice advocate, Davenport's response is nevertheless detailed and uncompromising on the specific issues Hahnel raises as he outlines the basis on which system-critical activists should decide what tactics and programmes they adopt. Davenport's article concedes that revolutionaries operating within the context of capitalist societies are engaged in struggles for reforms most of the time. Echoing Williams (2013), Davenport reiterates that despite the limitations of being embedded in capitalism, '[t]he question is how to struggle for reforms in a revolutionary way'. This question leads Davenport to distinguish between reformist tactics 'based on *accommodation to the class enemy*' (emphasis in original) and class-independent revolutionary tactics.

Davenport (2014) critiques reformist tactics such the one proposed by Hahnel because they constitute capitalist policy solutions, which it is not the task of the Left to work towards. He points out that while Hahnel is not the only academic of the Left to call for a 'shift towards reformist politics' because of the urgent need to take action on mitigating further climate change, '[i]t is bourgeois legislators' job to figure out how to implement … concessions within the framework of bourgeois policy – we should not do their jobs for them'. Davenport argues that supporting capitalist policy solutions such as carbon trading mechanisms entrenches capitalist relations of production rather than leading to the qualitative changes required to create sustainable and democratic, socially just societies. Support for such measures also limits the political imagination of the Left 'to what's possible within the capitalist system and pull[s] our politics rightwards' rather than affirming the message that 'another world is possible'. In addition, reformist tactics promote the view that 'progressive bourgeois legislators' are allies, and that the capitalist state is a neutral agent that can be negotiated with, rather than demonstrating that change will only come from confronting 'the state and the capitalist class'.

Most importantly, Davenport argues, such policies are (contrary to their claims) unrealistic; the actual implementation of carbon trading markets has not only failed to reduce GHG emissions – it has also enriched major polluters, and this is an inevitable outcome in a capitalist system. As Davenport (2014) notes,

[Hahnel] does not water down his proposal in order to make it palatable to the capitalist class and the politicians that serve them. He insists that a global cap-and-trade scheme would have mandatory emissions caps for all countries based on science, differential emissions caps based on principles of global justice, and enforcement mechanisms to prevent cheating and bogus carbon credits.

(Davenport, 2014)

However, Davenport also points out that it is precisely because this is a 'principled proposal' that 'The ruling class would never agree … [since this] would amount to voluntarily abandoning imperialism and agreeing to destroy huge sectors of highly profitable capital'. In summary, Davenport argues that 'it is unrealistic to expect capitalist governments to implement carbon trading schemes in accordance with … [Hahnel's] principles given that their purpose is to appear as if they are doing something while in fact they defend the continuation of business as usual for as long as possible'. Ultimately, he concludes, reformist tactics misdiagnose the causes of the ecological crisis: 'the climate crisis and other ecological problems are rooted in the structure [of the] capitalist system (not just a few sectors of capital, like the fossil-fuel industry, but the system itself), and cannot be reformed away'.

In contrast to reformist tactics that support capitalist solutions, Davenport argues that revolutionary tactics attempting to deal with the issues of the environmental degradation and climate crisis caused by capitalism can be developed with a more nuanced understanding 'of the *dynamics of power* in capitalism and socialism' (emphasis added):

> although in a capitalist system the capitalist class is always in power, at no point is the relationship of forces between classes static. The working class, the capitalist class, and other social forces are always vying for power and influence within the system and its units – within workplaces, neighbourhoods, and various levels of government, for example. Reforms, in general, happen when the working class or oppressed groups gain enough power to force the hand of the ruling class – putting them in the position of either implementing the reform or losing needed credibility and influence.

(Davenport, 2014)

It is this understanding of power relations, and of the importance of class struggle, that informs the development and implementation of revolutionary tactics that are, first and foremost, 'class-independent' in that they 'build popular power independently from state institutions' and are thus capable of 'confront[ing] the state and the capitalist class'. Davenport's distinctions between reformist and revolutionary tactics are summarised in Table 6.2.

Table 6.2 Reformist versus revolutionary tactics

Reformist tactics	Revolutionary tactics
Do not challenge hegemonic capitalist ideology	Aim to change participants' consciousness and encourage activists 'to think and act outside the confines of capitalist ideology'
Entrench capitalist relations of production	Lead to qualitative changes required to create sustainable and democratic, socially just societies
Limit the political imagination of the Left to what is possible within a capitalist system	Affirm the message that another world is possible; 'promote solidarity and draw attention to the conflict between the movement and the capitalist system'
Promote the view that reformist legislators are allies	[Aim to be], first and foremost, 'class independent'; 'build popular power independently from state institutions' and are thus capable of 'confront[ing] the state and the capitalist class'
Promote the view that the capitalist state is a neutral agent that can be negotiated with	Demonstrate that change can only come from confronting the state and the capitalist class
Promote what are erroneously perceived as 'realistic' solutions	Recognise that mainstream solutions are false: they do not, in fact, reduce GHG emissions but enrich major polluters instead; this is a fundamental feature of the capitalist system
Misdiagnose the root of the ecological crisis, which lies within the structure of capitalist relations, not just in some economic sectors	In the case of mitigating anthropogenic global warming, 'uncompromisingly demand a shift away from fossil fuels, done in a way that meets human needs'
See capital's power as hegemonic and immovable; do not understand the class struggle	Have a nuanced understanding of the dynamics of power in capitalism and socialism, recognising that 'reforms, in general, happen when the working class or oppressed groups gain enough power to force the hand of the ruling class…'
Solutions are individualistic	Encourage people to think in collective rather than in individual terms and focus on principles that unite working-class people
Solutions ignore the needs of the working class and subaltern groups	Address the 'immediate concerns of working-class people' while making connections between workers' struggles and struggles to counter 'ecological destruction and the overcoming of racial, national, and gender- and sexuality-based oppressions'
Solutions do not challenge existing power relations	Typically argue 'for stances that are further to the left of those in the mainstream of the movement, while posing them in a way that is relatable and responsive to the movement's aims'

Source: Adapted from Davenport (2014).

Given his argument that supporting carbon trading lends legitimacy to reformist policies that fail (and that ultimately *must* fail) to achieve their stated goal of reducing GHG emissions, Davenport makes a compelling case that the radical climate justice movement should not support this policy or any other market mechanisms proposed as solutions to anthropogenic global warming. This position, however, raises interesting questions about whether it is *ever* progressive to support reforms – an issue that crucially relates to questions about the labour movement (refer to Questions 1 and 3 in Table 6.1) and that the Left faced during the 2015 US oil workers' strike.

Case study 2: ecosocialist tactics – the question of class struggle

In the arguments he presents against Hahnel's proposal, Davenport (2014) also makes the important point that revolutionary strategy is necessarily *situational*, and that a programme that is reformist in one situation may be revolutionary in another. Such nuances are significant when analysing specific tactical decisions such as the decision by SCNCC members to support the 2015 US oil refinery workers' strike. This strike constituted the first major American oil worker stoppage since 1980, and had been called by the United Steelworkers Union (USW) on 1 February 2015 (Lusanne, 2015). The USW called the strike action after Royal Dutch Shell, as the lead representative of the petroleum industry in negotiations for a new three-year labour agreement, refused to concede to worker demands for 'better wages, lower health care costs, improved safety, and reduced hiring of temporary contract workers' (White, 2015). Socialist critiques of the union leadership's decisions during this strike point to its refusal to agree to rank-and-file calls for all 30,000 USW workers to go on strike (ICFI, 2015) and its insistence that the workers' main demands revolved around safety issues rather than wage increases (Bannon and White, 2015). These critiques align with renowned Marxist industrial relations scholar Richard Hyman's (1975) observation that trade unions have traditionally been problematic institutions for Marxists.

Originating in Britain as self-organised alliances of craft tradespeople who banded together to defend their rights as workers, trade unions were transformed (frequently forcibly) into compliant institutions that are routinely used by management (with support from both the state and often the union leadership) to control the working class. From their inception, trade unions have inhabited a terrain that is fraught with contradictions arising from their orientation as workers' collectives operating within the context of a system of capitalist relations that is implicitly accepted by the trade unions and the vast majority of the workers that constitute their membership. This fundamental contradiction has many additional contradictory flow-on effects; for example, while trade unions are, by their nature, situated in a conflictual relationship with capital (workers can only make wage gains or improvements to their working conditions at the expense of capital), the membership's dependence on having work if they are to survive forces workers (through their defensive organisations, the trade unions) to negotiate with employers and cooperate in their own continued exploitation. However, while trade unions are undoubtedly institutions that are firmly entrenched within the capitalist system in the advanced capitalist economies, and have undergone processes that have rendered them organisations that are used by the ruling class and its managers (including trade union officials) to control labour, they are nevertheless working-class organisations whose members cannot be ignored by any counter-hegemonic movement that wants to develop broad-based support for its project.

The ecosocialist argument that it is important to support workers' struggles as part of the fight against climate change is merely an extension of the ideas of Marx, Engels, Lenin, Gramsci and other Marxists that the working class is the key agent of social change within a capitalist system.[7] Ecosocialist efforts to ally with workers, as evidenced in their decision to support the 2015 oil workers' strike, is especially important in periods of organic crisis; as Hyman argues:

> In periods of unrest and instability, the presence in positions of influence of workers with a developed oppositional ideology can be of immense significance. When engaged in collective struggle, workers are most susceptible to the appeal of new world-views: the 'deviant' elements in working-class attitudes are thrust to the fore, while the conventional assumptions of 'official' society momentarily lose their hold.
>
> (Hyman, 1975, pp. 176–177)

It is not surprising, then, that one of SCNCC's key demands is to: 'Provide full employment, transitioning millions from military and fossil-fuel related jobs to union jobs creating a renewable energy infrastructure. In the fight to save the environment, working people must not be left behind' (SCNCC, n.d., p. 9). Ecosocialist strategy involves fighting for both short- and medium-term reforms with a long-term aim of fundamental system change, and the reformist measure calling for full employment is justified as follows.

> In any case, we can never expect workers to support the changes needed to save the planet and their children unless they are provided with other jobs, at least as good with at least equivalent pay and benefits, as the ones they are losing.
>
> (SCNCC, n.d., p. 9)

However, given the aims of the climate justice movement to shut down the fossil fuel industry and demand a transition to the use of renewable energy, some climate justice activists questioned the tactic of supporting the USW oil workers' strike (Johnson, 2015; Rugh, 2015).

While support for striking workers employed in the fossil fuel industry may appear incongruous because it seems to contradict ecosocialist demands to 'uncompromisingly … shift away from fossil fuels' (Davenport, 2014),

several climate justice movement actors presented strong arguments for supporting this industrial action. Citing historian and labour activist Jeremy
Brecher, for instance, Trish Kahle (2014) argues that ecosocialists need to
'break down the false "jobs versus environment" dichotomy' and act on the
understanding that 'the exploitation of workers and the degradation of the
environment go hand in hand' so that workers and environmentalists must
'evolve toward a common program and a common vision'. Kahle further
argues that it is particularly important for climate justice activists to engage
with workers employed in energy industries because these workers 'occupy
a special place at the nexus of capitalism's ecological destruction and human
exploitation, a place that is simultaneously powerful and vulnerable'.

Kahle argues that the position that energy workers occupy in a capitalist
system is:

> [p]owerful, because energy sets the entire economic system in motion,
> and any action taken by the workers responsible for producing this
> energy quickly fans out to every sector of the economy. And vulnerable
> because the radical realignment of energy production they have the
> power to affect can threaten their livelihoods.
>
> (Kahle, 2014)

Workers in the energy industry thus 'have the power to bring our planet
back from the tipping point', although this can only be achieved if energy
workers regain control of their organisations from 'a labor leadership that
appears to be willing to risk the future of our species for slightly relieved
unemployment' (Kahle, 2014). In this respect, Kahle (2015) argued that
ecosocialist support for the 2015 oil workers' strike demanding safer working
conditions was crucial.

> [F]or ecosocialists, it's not just the immediate demands of workers that
> are important, but the long-term ramifications of a victory. Workers
> have power if they act collectively. Just as they can stop oil production
> (30,000 workers have the capacity to halt 64 percent of the nation's
> refining capacity), they can halt capitalism's assault on the planet. To be
> sure, there's not a direct line between a strike for better working con-
> ditions and a strike for new energy and a more just economy. It is the
> job of ecosocialists to demonstrate the deep connection between the
> two, to offer an analysis and strategy of struggle that speaks to workers'
> lived experiences.
>
> (Kahle, 2015)

Referring to the oil workers' strike as containing within it 'the seeds of an
alternative: class-struggle environmentalism', Kahle suggested several concrete
actions that climate justice activists could engage in to support the oil
workers' action, including donating to strike funds, participating in picket

lines and inviting workers to talk about the strike at community meetings and on campuses. Such support would have both a short-term goal of helping workers win the strike but also a longer-term goal of helping unions to become stronger and perhaps even join in the struggle to mitigate climate change and environmental destruction. The SCNCC website provided links to Kahle's articles, and SCNCC members joined picket lines in support of the striking workers (Rugh, 2015). Several other environmental, trade union and Left-leaning groups such as the California Nurses Association, the Labor Network for Sustainability, 350.org, the Sierra Club and Oil Change International also expressed their support for what developed into the largest oil refinery workers' strike in 35 years (Johnson, 2015; Light, 2015; Rugh, 2015; Turnbull, 2015; Uehlein, 2015).

While the SCNCC (n.d., p. 9) demand for 'full employment, transitioning millions from military and fossil-fuel related jobs to union jobs creating a renewable energy infrastructure' is, arguably, better served by confining its collaborations to other trade union organisations that better align with this aim – such as Trade Unions for Energy Democracy (TUED, n.d.) and the International Transport Workers' Federation (ITF, n.d.; see also Felli, 2014), confining actions to terrains on which everyone already agrees does not extend the climate justice movement, and the challenge is to create a broad popular movement that is powerful enough to fight for a habitable planet. Moreover, an application of Davenport's guidelines for determining whether a course of action is progressive or not demonstrates that SCNCC support of the oil workers' strike was overwhelmingly progressive. By participating in picket lines, SCNCC members not only concretely demonstrated a nuanced understanding of the dynamics of power relations in capitalist societies as well as their solidarity with the workers and their concern for workers' wellbeing and safety, but they also created possibilities for further collaborations between ecosocialists, workers and communities by demonstrating that such broad coalitions are possible.

Supporting the oil workers enabled SCNCC members to participate in a collective action that by its nature challenged the power of the fossil fuel industry (albeit to a limited extent) and contributed towards efforts to strengthen future challenges to this power. As Kahle argues, the power of the fossil fuel industry needs to be challenged on many different fronts.

> As significant as winning better working conditions would be – for the workers themselves, as well as for the communities around the refineries – it's also important to consider what the long-term effects of this strike could be in terms of increasing militancy and organization. Strong unions can fight for jobs and the planet, but we still have a long way to go. To defeat the oil industry ecologically, we need to defeat it ideologically, politically, socially, economically: this strike is a step in that direction, and that's why it's so important to win.
>
> (Kahle, 2015)

In addition to these arguments for supporting workers in the fossil fuel industry in their challenges to capital, while more overtly progressive campaigns such as the UK Campaign Against Climate Change's call for 'One Million Climate Jobs' funded by the public sector (Empson, 2010) are undoubtedly a great step forward in the fight for a 'just transition' to a sustainable energy system, they are not completely unproblematic as they exist in the context of an emerging focus on a so-called 'blue–green' alliance between 'blue-collar' workers and the 'green' (environmental) movement. Such blue–green alliances (and related notions of a 'green economy', 'green technology', 'green jobs' and 'sustainable development') can be easily coopted into supporting, legitimising and entrenching liberal and neoliberal agendas to create new avenues for capital accumulation at the expense of both nature and subordinate classes and groups. Returning to de Lucia's distinction between 'system-critical' and 'system-friendly' approaches, the current conjuncture is particularly perilous for radical climate justice activists seeking real solutions to the organic crisis given the struggle within the capitalist class between the faction supporting the long-established vested interests of fossil fuels and the faction supporting an emerging 'green capitalism', which could gain much-needed leverage if it succeeds in mobilising a broad support base comprised of subordinate classes and groups to build a re-formed historical bloc legitimising capital's continued expansion.

The new 'green economy' thus constitutes an arena of struggle between those forces working towards building a new hegemony and those trying to build a counter-hegemonic movement. This battle is being waged in both *discursive* terms (in a war of position) involving the interpretations of contested concepts such as 'sustainable development', 'green economy', 'green jobs', 'climate justice', a 'just transition' and 'system change' and on the terrain of political action (which could be described in terms of a limited 'war of manoeuvre' with respect to the development of appropriate strategies and tactics). The war of position is crucial as some interpretations of the contested concepts could result in a 'reformed green capitalism' that furthers the neoliberal agenda of capital accumulation without, in the worst-case scenario, even actually making the changes required to mitigate AGW and general environmental degradation.[8] This could open the path to an intensified assault on, and the deeper economic exploitation of, both nature and subordinate classes.

The 'green capitalism' route is also highly likely to lead to dangerous (but very profitable) large-scale projects such as attempts to 'sequester' carbon and, even more dangerously, large-scale geo-engineering projects.[9] Thus, while intellectuals of the Left argue that capitalism is incapable of addressing climate change and see 'saving the planet' as a catalyst around which to organise in order to achieve a more socially just world order, the need to address climate change could equally be the uniting issue that justifies and legitimises the power of global capital (at least until the planet is no longer habitable) while, as Goodman and Salleh (2013) note, simultaneously opening up much-needed avenues of new 'economic growth' and capital accumulation (see also Boehnert, 2015; Burkett, 2014; and Elgot, 2015). Moreover, de Lucia (2014, p. 68) points out that at the same time as rejuvenating the profitability of global capitalism, the concerted attempt to present market mechanisms and technological fixes as the only possible solutions to the climate crisis simultaneously deflects more radical anti-capitalist critiques. De Lucia concludes that 'antagonistic forces are thus either embraced as potential partners or ejected from the realm of the reasonable', and it is thus that *trasformismo* is attempted (*ibid.*). The task of the TCC and its allies is therefore to achieve consensus on the notion that only 'green capitalism' and its 'green innovation' and 'green technology' can 'save' the planet, and there are factions within capital that clearly recognise this and are working towards this end (Goodman and Salleh, 2013).[10] The UNFCCC's annual COPs constitute an important forum in which the project to promote 'green capitalism' is advanced, and ecosocialist contributions to discussions amongst climate justice activists who were planning actions for the 2015 COP-21 in Paris demonstrate a profound awareness of this issue.

Case study 3: ecosocialist tactics – shutting down COP-21

In the lead-up to COP-21 in Paris in 2015, ecosocialists participated in several discussions about possible actions that could be taken by climate justice activists given the likely official outcomes of the meeting and the prevailing balance of forces.[11] Anticipating that any agreement emanating from COP-21 would, like previous official outcomes, promote the implementation of 'false solutions' such as the extension of carbon markets and the further commodification of nature (Carbon Trade Watch, 2015; Temper and Gilbertson, 2015), radical climate justice activists debated how best to approach these climate talks in a way that avoided both legitimising the official negotiations and making useless appeals to authorities to adopt more effective and socially just policies. As the editors of the Carbon Trade Watch publication discussing the forthcoming COP-21 put it:

> So what can we do? We can start by recognizing that the climate negotiations are making things worse. We need to think beyond the UNFCCC and to stand in active solidarity with those who are at the frontlines of fighting the climate and environmental criminals while defending their territories.
>
> (Carbon Trade Watch, 2015, p. 5)

Coalition Climat 21, the coalition of approximately 130 activist organisations established to coordinate activities in France for the duration of COP-21, held a strategy meeting in Tunis on 23 and 24 March 2015 (Bond, 2015, p. 22). Ecosocialist Patrick Bond (*ibid.*, pp. 23–27) critiqued the prevailing approaches that sacrificed political analysis in their attempt to seek unity at this strategising meeting, and argued that a more appropriate strategy would be that proposed by Pat Mooney to 'shut down' the official COP, which many radical climate justice activists refer to as a 'Conference of Polluters', on the grounds that 'no deal is better than a bad deal'. In a series of email exchanges that occurred between March and June 2015 and that was compiled and distributed by SCNCC member John Foran (2015), ecosocialists continued to discuss the option of shutting down the official climate talks. Most of the participants in these discussions supported this action, but some raised concerns about the ability to succeed given that the French state would be out in force to pre-empt such actions and also because it was unlikely that a broad enough consensus within the climate movement would agree to this strategy. Another concern raised was that if it acted on such an aim, a failure to shut down the official talks would have a demoralising effect on the broader climate justice movement.

While Coalition Climat 21 ultimately decided not to attempt to shut down the official talks, a variety of other mass actions were planned with the aim of building a stronger climate movement that would 'move through' Paris2015. Referring to 'Paris2015' was symbolically important because, in rejecting the label 'COP-21', activists emphasised that the climate justice movement had convened in Paris not to lend any legitimacy to the official negotiations but to build their independent climate movement in order to continue local struggles for climate justice after the anticipated failure of COP-21 (Temper and Gilberston, 2015). The major planned mass actions for Paris2015 adopted the theme that 'red lines are not for crossing' and Coalition Climat 21 decided that the main mass action, the 'December 12' ('D12') red line protest, would occur at the end of the talks (SCNCC, 2015). By holding the main action at the end of the talks, the climate justice movement would have 'the last word'; instead of making more futile calls for climate justice that official negotiators would ignore (thus demoralising activists), the climate justice movement would instead denounce the ineffective and damaging decisions of the official COP-21 delegates after they reached the agreement that was predicted to 'burn the planet' and, as climate justice activist Pablo Salon cautioned, was likely to result in 'setting a course for geoengineering' (Foran, 2015; SCNCC, 2015).

The activist plan was to introduce five 'red lines' that should not be crossed (because they represent 'minimal necessities for a liveable planet') in mass marches on the eve of the COP-21 opening, and to have the final word at the end of the official negotiations.

> [W]hen the deal inevitably crosses these red lines, people encircle the summit and in a show of collective power and shaming, refuse to let the delegates return home to carry out the criminal agreement in their communities.
>
> (SCNCC, 2015)[12]

In addition to these major mass-protest actions at the beginning and end of the official talks, Coalition Climat 21 also planned to facilitate and host a series of other events around the city of Paris, including: the decentralised Climate Games involving acts of nonviolent civil disobedience by small, self-organised groups participating in acts of 'creative resistance' (see McDonald, 2015); daily presentations by social movement activists and a variety of civil society groups at the alternative Climate Action Zone during the second week of the official negotiations; and many stalls, events and discussions at a two-day alternative climate summit, Alternatiba, held on 5 and 6 December in Montreuil, a working-class Parisian suburb (SCNCC, 2015).

When the French government declared a state of emergency in the aftermath of the 13 November 2015 Paris terror attacks and banned all planned climate movement mass mobilisations, Coalition Climat 21 organised alternative actions (such as the 'human chain' of 29 November 2015 along the march's original planned trajectory). Moreover, displaying a dedication that suggests that calls to try to shut down the official talks may not have been as unpopular as Coalition Climat 21 organisers had feared, many activists defied the government ban on mass protests so that the D12 actions went ahead (as did many of the other non-violent civil disobedience actions) even in the face of increased police repression (de Moor, 2015). This spontaneous show of defiance by climate movement activists, coupled with the realities of the rapidly unfolding planetary-scale ecological catastrophe that climate change is just a part of, suggests that future ecosocialist arguments for less accommodative tactics may garner more widespread support within the climate movement. The evaluation of ecosocialist strengths and weaknesses is therefore both partial (based on the limited number of relevant issues analysed) and necessarily transient (based on a fast-changing 'present', while we live in an age defined by an organic crisis where the balance of forces can shift very rapidly and dramatically as events unfold).

Ecosocialist strengths and challenges: a partial and transient evaluation

The fact that the ecosocialist suggestions to try to shut down the COP-21 negotiations were not adopted by Coalition Climat 21 demonstrates that the ecosocialist contingent's influence within the leadership of the broader climate justice movement is limited; it does not, however, necessarily demonstrate that ecosocialists have no influence at all in the broader climate justice movement. The red lines mass action that was planned for the end of the Paris 2015 climate negotiations was meant to send a strong message that the climate justice movement was dissatisfied with the official outcomes of COP-21, and the decision to take this action was perhaps at least partially motivated by ecosocialist inputs in the discussions about appropriate strategies and tactics.

Moreover, while ecosocialist calls for more radical actions at Paris2015 were not adopted, this did not lead to sectarian divisions and ecosocialists continued to circulate information about the actions that were planned, with some ecosocialists also attending Paris2015 and participating in these actions.

This demonstrates that ecosocialists adopt a disciplined approach that prioritises solidarity and tries to avoid traditional 'Left' sectarianism as they participate in the struggles for climate justice, even if their preferences for stronger actions are not adopted. While ecosocialist ideas about appropriate tactics for building a counter-hegemonic climate justice movement have not yet been widely adopted, their understanding of the kinds of oppositional tactics that are necessary if they are to achieve the aim of furthering the project of building a post-capitalist ecological socialist society is well developed.

Social facts and competing ecosocialist ideas

One of the greatest strengths of ecosocialist theorists and activists is their powerful and clear analysis of the causes of capital's organic crisis, including the current climate crisis. As discussed in Chapters 3 and 5, third-stage ecosocialists apply classical Marxism's analytical tools in order to explain why system change is necessary, and many of these ideas are now commonly discussed amongst activists within the climate justice movement. To the extent that ecosocialist ideas have become more prominent, they have replaced the previously prevailing 'social facts' that 'there is no alternative' to capitalism (to quote one of Margaret Thatcher's famous sayings). Ecosocialists' familiarity with the Marxist notion of the dialectic has also enabled them to develop a sound understanding of the profound implications of the Anthropocene.

To Marxists, there is no difficulty in understanding the Anthropocene as a 'phase shift', whereby quantitative changes to natural systems and cycles caused by the normal operations of capital have reached a point where they have cumulatively resulted in a qualitative change in the Earth System. It is therefore not surprising that it was Ian Angus' book *Facing the Anthropocene: Fossil Capitalism and the Crisis of the Earth System* (2016), that Clive Hamilton (2017) singled out to praise for the clarity with which it explains the implications of this new geological epoch. Angus' focus on the Anthropocene is very valuable, and the concept and its implications are now widely discussed amongst climate justice movement activists and are also gaining a wider audience amongst the readership of articles on ecosocialist websites. Despite evidence that ecosocialist ideas are becoming more prevalent within the climate justice movement, however, it would be false to conclude that they have made inroads within the broader society, where 'social facts' generally continue to prevail (although there is a growing 'common sense' view that there are widespread problems with the current system). On the other hand, ecosocialist writings have some influence, as is evidenced by many of the ideas prevailing amongst climate movement activists.

Revisiting ecosocialist tactics: what role 'prefiguration'?

While the first and second case studies analysed in this chapter demonstrate that ecosocialists have a sophisticated understanding of the sorts of oppositional tactics that are appropriate for furthering their long-term goals, the

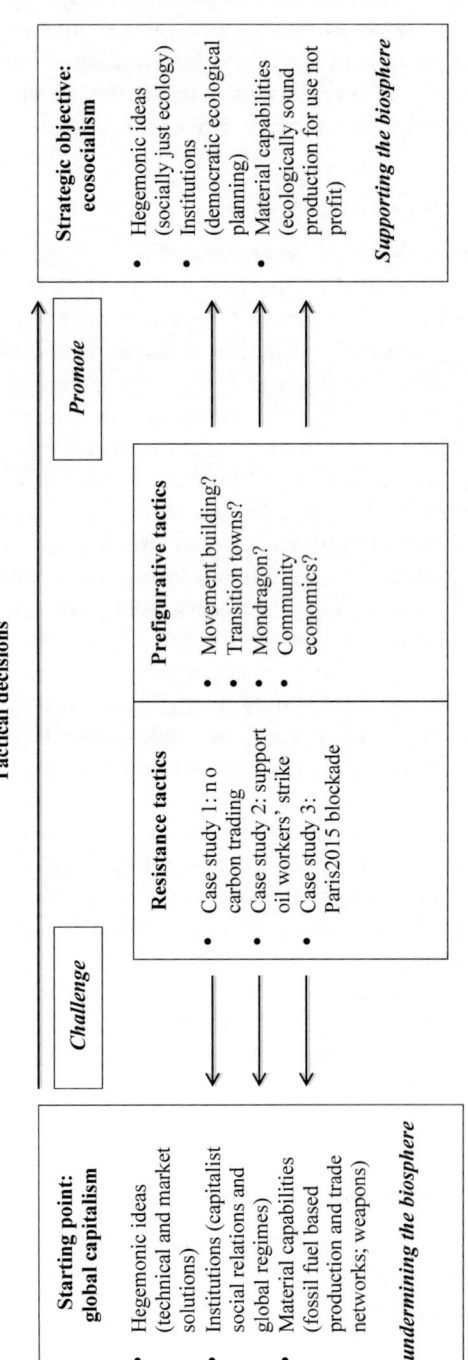

Figure 6.1 Ecosocialist strategy and tactics.

third case study illustrates that these tactics do not necessarily garner support within the wider climate justice movement. Ecosocialist suggestions to 'shut down' the official COP-21 climate negotiations are an instance of what Naomi Klein (2014) refers to as 'Blockadia' – a nonviolent civil disobedience act that blocks actions that damage the planet. Like earlier SMOs, such as the US 'Movement for a New Society' that was active in the 1970s and adopted the strategy 'oppose and propose' (Cornell, 2011), in addition to the more system-critical 'Blockadia' actions it planned, Coalition Climat 21 also organised the establishment of Alternatiba, a 'village of alternatives' where the climate movement held its own climate summit. Ecosocialists seem to find the idea of prefigurative projects such as Alternatiba problematic, however, and it is perhaps a potentially important tactic that deserves more discussion and debate. Figure 6.1 provides a schematic summary of tactics ecosocialists could adopt. It depicts a range of tactical decisions, from those that challenge the current hegemonic bloc (refer to the left-hand side of the figure), to examples of tactics that could promote the building of new institutions within the current disintegrating world order (refer to the right-hand side of the figure).

While the role that prefigurative projects could play in ecosocialist politics was raised in Solidarity's discussion about strategies and tactics (refer to Table 6.1, Question 6), as noted previously it is not an issue that generates much discussion amongst ecosocialists. Most ecosocialists affirm the need for *visions* of an alternative to capitalism (an issue that is understood in both strategic and tactical terms); however, the extent to which ecosocialists should participate in prefigurative actions and projects that try to build alternatives in the present is contested. Most ecosocialist understandings seem to narrowly limit the idea of prefiguration to individualistic, personal 'lifestyle' changes such as riding bicycles, adopting vegetarian or vegan eating habits and changing lightbulbs. On the occasions that the ecosocialists interviewed were asked about their views of the role prefiguration could play in effecting system change, their response was always that while they have nothing against such personal lifestyle changes, people adopting these individual practices should recognise that they will not, in and of themselves, achieve the social transformation necessary to maintain a habitable planet. Limiting the definition of prefiguration to encompass only changes in individual behaviour is questionable, however, because there are other examples of *collective* prefigurative projects that could be both educational and help to create nascent communities for future collective action. While not without their problems and challenges, examples of collective prefigurative projects that may provide valuable insights to inform ecosocialist tactics include the Transition Town movement and the European Alternatiba projects and events organised by various climate movement actors.

The Transition Town movement

Originally pursued because its founders were concerned that peak oil, climate change and economic instability would create conditions forcing people to change how they live, the Transition Town movement aims to create 'resilient communities' by encouraging the people living in their towns and cities to transition 'to more sustainable lifestyles around localisation, co-operation, organic food production and energy descent into renewable and reduced energy systems' (Weston, 2014, pp. 185–186). While the Transition Town movement is not without its problems (with the notion of 'resilience' being a contested term which is largely used to justify adaptation practices rather than pursuing the large-scale systemic changes necessary to mitigate climate change and environmental destruction), it nevertheless presents a model that could form the basis of alternative collective prefigurative projects.[13] For example, ecosocialist political theorist Del Weston (2014, p. 183) argues that:

> Without a clear way forward, I believe that we must validate and support all attempts to break from the capitalist social relations of production and the imperatives of capitalism and work towards solidarity amongst alternatives. By doing so, by embracing and supporting the Transition Towns, the Abalimi Bezekhavas, the Campesino movements, we are building a solidarity across alternatives to capitalism, showing examples of how it is possible for people to become empowered and positively involved in making their own history, that it is possible to form new values in which we are comfortable and at home ... We must at every opportunity show that capitalism is not the only viable political and economic system, that there are coherent and current alternatives to it.[14]

The Transition Town movement has the decided advantage that it works within local communities that remain integrated within the broader society, unlike other groups such as the The Bruderhof, the religious prefigurative intentional community that Joel Kovel (2007) discusses as a model that ecosocialists can draw lessons from. Members of this collective problematically seem to live in isolated communities that are largely divorced from the wider society, and many ecosocialists are justifiably sceptical of the utility of such projects that entail withdrawing from mainstream society.[15] This is because, while isolated communities may find it easier to adopt sustainable living practices, they are unlikely to attract 'ordinary people' to participate in collective actions that promote solidarity and develop new ways of thinking and doing things. It seems logical that only prefigurative projects that are located within, and welcome the participation of, the wider community would have any chance of furthering the aim of building a new society. There are many arguments to support the view that properly developed collective prefigurative projects are not only valuable but also necessary at this point in time, one of which being their indispensible regenerative function at a time when climate movement actors are experiencing much stress. The value of participating in activities with others in a prefigurative 'alternative village' became evident to me when I attended the Paris2015 events, and I recount those experiences as an illustration of the regenerative function of such projects.

Alternatiba

Alternatiba was a temporary Paris2015 project that showcased many ideas of socially just solutions to climate change while simultaneously providing an important space within which climate movement activists could experience a sociality and conviviality that was completely absent from the 'Climate Generations Areas' set aside for 'civil society' at the official COP-21 talks. Attending this two-day event, which provided an enjoyable (albeit temporary) prefigurative experience at a time that was exceedingly stressful given the tensions generated by the French state's decision to declare a state of emergency and ban planned climate movement actions, was both motivating and restored my conviction that people can, indeed, achieve great things when they cooperate. I had found the previous days' experiences at Paris2015 very stressful. On my first day in Paris, I was halfway up the stairs leading from the Metro to the Place de la République to join other Paris2015 climate movement participants when the Metro tunnel filled with teargas. The situation was chaotic, with people running down the stairs to escape police who were chasing them colliding with commuters who had just disembarked from the train, and with everyone coughing and rubbing their streaming eyes, which were burning as a result of the teargas.

While my visits to the 'Climate Generations Areas' in Le Bourget, where the official COP-21 negotiations were taking place, were not as dramatic, they were not pleasant experiences. Members of 'civil society' disembarking from the trains had to catch designated buses to get to the venue and, once there, were treated suspiciously and subjected to a security check as thorough as those conducted at airports. Inside the sterile 'Climate Generations Areas' there were few food and beverage stalls, and people had to wait in long queues to buy the limited range of food and beverages available (with many products, including coffee, being sold out early in the day). In contrast, a friend of mine who had registered to attend the official talks as a representative of an NGO said that there was no shortage of free and excellent food in the areas where the negotiations were taking place (and to which the members of 'civil society' were not allowed entry). This seemed to me a microcosm of the inequality and injustice that characterises all of life in a world dominated by global capital, and perhaps it was the contrast between the official 'Climate Generations Areas' and the Alternatiba experience that made such a deep impression on me when I arrived at the 'village of alternatives' in Montreuil, a working-class suburb of Paris.

Guardedly emerging from the Montreuil Metro, my heart sank when someone wearing an official-looking fluorescent vest approached me. To my relief and delight, the person who approached me turned out to be a volunteer working with Coalition Climat 21 who, rather than teargassing me, smiled broadly and welcomed me, handing me a map and pointing out the location of various stalls and facilities. There were no security checks and no police in sight; instead, there was a relaxed, festive atmosphere with many stalls selling locally produced foods and beverages. The food was not only abundant, but it was also very fresh, healthy and reasonably priced. As well as people who were clearly climate activists, there were also many families with babies and young children and a seemingly wide spectrum of people from all walks of life. In addition to the festivities and stalls, there were a number of events one could attend, such as a presentation by the TUE and another by the UK group advocating 'One Million Climate Jobs'. Alternatiba's value lay in how it incorporated interesting learning experiences while simultaneously promoting a general atmosphere of convivial solidarity and also demonstrating what can be achieved by people working collectively and cooperatively despite their limited material resources. Another important feature of Alternatiba during Paris2015 was the way in which it was embedded in the local community of Montreuil, attracting local residents and inviting them to join in the organised activities.

My experiences at the Paris2015 Alternatiba event reinforced my interest in the value of prefigurative projects to supplement (not replace) direct actions aimed at challenging existing power structures. Ecosocialists participating in suitable community projects could engage in discussions that build on participants' 'common sense' while simultaneously building solidarity and prefiguring the kind of social relations they want to build. Such projects may also provide the resources that people will need (particularly with respect to practices of solidarity) as the effects of global warming continue to unfold in 'extreme weather'-related disasters that will force people to rely on their own resources because governments will be unlikely to respond to them effectively – particularly when these disasters affect the most disadvantaged people, who are perceived as 'disposable' and have no political power, as happened in the aftermaths of Hurricanes Katrina and Sandy in the US and as is happening at the time of this writing, with Hurricane Irma unleashing its force in the US and the Caribbean.[16]

Notes

1 Refer to Löwy (2015, pp. 77–82) for a copy of the *First Ecosocialist Manifesto*.
2 The annual WSF was established by social movement activists in 2001, and is described as a 'dynamic process' that 'operates autonomously from the interstate system and has become the leading focal point for transnational mobilization and interchange among movements'; it aims 'to build transnational and cross-sectoral movement alliances and encourages ongoing efforts to link local struggles with a critique of the global neoliberal economic order' (Smith and Wiest, 2012, p. 2).
3 Prominent ecosocialists advertised to speak at this conference included Joel Kovel, Fred Magdoff, Brian Tokar, Chris Williams, Richard Smith and John Riddel (Ecosocialist Conference, 2013). Other prominent speakers included labour activists and authors Sean Sweeney and Jeremy Brecher, and US Green Party nominee for the 2012 and 2016 presidential campaigns Jill Stein (*ibid.*).
4 There is not always a clear demarcation between strategic and tactical questions, however; for example, Question 2 in Table 6.1 has both strategic and tactical elements.

5 As Foster and Clark (2015) explain at the beginning of the article, 'crossing the river of fire' refers to William Morris's metaphor for becoming 'a critic of capital as a system'.

6 Davenport was only one of many to respond to Hahnel's proposal; other responses included an online debate between Hahnel and well-known South Africa-based climate justice activist Patrick Bond that occurred on ZNet's debate webpage. Bond (2013) also responded to Hahnel's proposal in an article published in *Capitalism Nature Socialism*, as did environmental activist Gar Lipow (2014, 2015).

7 Notwithstanding the debates about whether or not trade union officials and privileged sections of the working class (the so-called 'labour aristocracy') act in the interests of capital and thereby wilfully betray the working class as a whole, more nuanced analyses of trade unionism identify underdeveloped class consciousness amongst the majority of the rank-and-file membership as the main reason explaining why workers are not more militant (for example, see Hyman, 1975; McIlroy, 2014).

8 Boehnert (2015) provides a brief overview of contested meanings of the 'green economy', with a useful summary of different perspectives on 'Economic approaches to the environment' in figure 2 in her article. Refer to Goodman and Salleh (2013) for a detailed discussion of the years of effort by a variety of UN institutions, business interests, CSOs and trade union groups to develop a global consensus around the idea of a shift towards 'green capitalism' to facilitate continued economic growth. McIntyre and Hillard (2012) also provide an important analysis of the way in which, contrary to common misperceptions, the original 'New Deal' in the US constituted a victory for capital. Their argument could be extended to contemporary efforts to achieve a 'new green deal', which are likely to have similar results.

9 Huttunen, Skytén and Hilden (2015) identify a clearly discernible trend of increasing attention being paid in recent academic literature to the development of large-scale geo-engineering 'governing regimes' and also to ways of gaining popular consent for more radical geo-engineering projects involving solar radiation management (SRM).

10 For example, refer to Faccer, Nahman and Audouin (2014) for a comprehensive summary of three contending discourses (which they label as 'incrementalist', 'reformist' and 'transformative') on how capitalism could respond to global warming.

11 Debates and decisions about appropriate climate justice movement strategies and tactics for COP-21 were published on a dedicated SCNCC website (SCNCC, 2015b) as well as in various activist documents (for example, Carbon Trade Watch, 2015; Foran, 2015; SCNCC, 2015a, 2015c; Temper and Gilbertson, 2015).

12 Suggestions for relevant red lines not to be crossed in the discussion paper circulated by SCNCC (2015c) included GHG emission reduction targets, equity, finance, justice and compliance.

13 Naomi Klein suggests substituting the term 'regenerative' for 'resilient' to avoid the latter's associations with passive processes that imply 'the ability to absorb blows and get back up' (Klein, 2014, p. 447).

14 Del Weston's reputation amongst ecosocialists is evident in that her book, *The Political Economy of Global Warming: The Terminal Crisis*, is listed by Ian Angus (2017) as one of the 'Essential Books on Marxism and Ecology' on the Climate & Capitalism website.

15 Kovel (2007, p. 211) emphasises that the *Bruderhof* model does not fully meet the requirements of an ecosocialist prefigurative community 'because an ecosocialist

society must be fully democratic, and not the province of any religious interpretation; and, more specifically, because the Bruderhof are not actually ecocentric in their orientation'.

16 Governor of Florida Rick Scott warned residents of the state that

> At some point, people are going to be on their own, so to speak, for a period of time during which the flooding and raining and the wind bear down on them, and they need to be prepared if they are in that path and haven't taken some action to get themselves in a less dangerous position.
>
> (Luscombe, Pilkington and Smith, 2017)

However, as Pilkington (2017) notes, some people he interviewed simply do not have the resources to follow Governor Scott's advice and 'get themselves in a less dangerous position'.

References

Angus, I. (2008). Comments on the Second Ecosocialist Manifesto. Climate & Capitalism, 15 June. Available at http://climateandcapitalism.com/2008/06/15/comments-on-the-second-ecosocialist-manifesto-and-some-key-issues-facing-ecosocialists-today/.

Angus, I. (2011). How to Make an Ecosocialist Revolution. green left, 7 October. Available at www.greenleft.org.au/content/how-make-ecosocialist-revolution-watch-video.

Angus, I. (ed.) (2012). Welcome to the new Climate & Capitalism. Climate & Capitalism, 12 March. Available at http://climateandcapitalism.com/contact-climate-and-capitalism.

Angus, I (2017). Essential Books on Marxism and Ecology. Climate & Capitalism, 10 November. Available at http://climateandcapitalism.com/2017/11/10/essential-books-on-marxism-and-ecology.

Angus, I., Riddell, J., Proyect, L. and Silverman, B. (2013). Ecosocialist Conference Shows Potential for a United Green Left in North America. Climate & Capitalism, 23 April. Available at http://climateandcapitalism.com/2013/04/23/ecosocialist-conferencemajor-advance-for-green-lefts/.

Bannon, E.P. and White, J. (2015). USW Rejects New Offer by Shell as Oil Strike Enters Sixth Day. World Socialist Web Site, International Committee of the Fourth International (ICFI). Available at www.wsws.org/en/articles/2015/02/06/oilw-f06.html.

Becker, M. (2014). Ecuador's Bitter Choice. *Against the Current* 168, January/February, 10–13.

Bloom, S. (2015). Six Questions for Ecosocialists: The Road Ahead, the Struggle at Hand (Response by Steve Bloom). Solidarity: A Socialist, Feminist, Anti-Racist Organization, 6 May. Available at www.solidarity-us.org/ecosocialistresponsebloom.

Boehnert, J. (2015). The Green Economy: Reconceptualizing the Natural Commons as Natural Capital. *Environmental Communication*, 10(4), 395–417.

Bond, P. (2013). Climate Crisis, Carbon Market Failure, and Market Booster Failure: A Reply to Robin Hahnel's 'Desperately Seeking Left Unity on International Climate Policy'. *Capitalism Nature Socialism*, 24(1), 54–61.

Bond, P. (2015). Challenges for the Climate Justice Movement: Connecting Dots, Linking Blockadia and Jumping Scale. In Temper, L. and Gilbertson, T. (eds), *Refocusing Resistance to Climate Justice: COPing In, COPing Out and Beyond Paris*. EJOLT Report, 23, 17–33.

Burkett, P. (2014). *Marx and Nature: A red and Green Perspective*. Chicago, IL: Haymarket Books.

Carbon Trade Watch (2015). Paths Beyond Paris: Movements, Action and Solidarity towards Climate Justice. Carbon Trade Watch, 1 December. Available at www.carbontradewatch.org/publications/paths-beyond-paris.html.

Climate & Capitalism (2007). Ecosocialist International Network (News Release). Climate & Capitalism, 9 October. Available at http://climateandcapitalism.com/2007/10/09/ecosocialist-international-network-news-release/.

Climate & Capitalism (2013). Join the Ecosocialist Contingent in Washington, February 2013. Climate & Capitalism, 10 February. Available at http://climateandcapitalism.com/2013/02/10/join-the-ecosocialist-contingent-in-washington-february-17/.

Cornell, A. (2011). *Oppose and Propose! Lessons from Movement for a New Society*. Oakland, CA: AK Press.

Davenport, N. (2014). A Response to Robin Hahnel's Open Letter to the Movement. New Politics, 4 March. Available at https://newpol.org/response-robin-hahnels-open-letter-movement/.

de Lucia, V. (2014). The Climate Justice Movement and the Hegemonic Discourse of Technology. In Dietz, M. and Garrelts, H. (eds), *Routledge Handbook of the Climate Change Movement*. New York: Routledge, 66–83.

de Moor, J. (2015). Climate Justice Activism under the 'State of Emergency'. In Brüggemann, M. (ed.), *Media Representations of Climate Change Politics at COP21: The End of the Beginning*. Available at https://climatematters.blogs.uni-hamburg.de/wp-content/uploads/2016/01/Watchblog.pdf.

Dietz, M. (2014). Debates and Conflicts in the Climate Movement. In Dietz, M. and Garrelts, H. (eds), *Routledge Handbook of the Climate Change Movement*. New York, Routledge, 292–307.

Ecosocialist Conference (2013). Ecosocialist Conference. Facebook event, 20 April. Available at www.facebook.com/events/174607756020503/.

Ecosocialist International Network (EIN) (2007). Ecosocialist International Network (News Release). Climate & Capitalism, 9 October. Available at http://climateandcapitalism.com/2007/10/09/ecosocialist-international-network-news-release/.

Ecosocialist International Network (EIN) (2009). The Belem Ecosocialist Declaration. In Angus I. (ed.), *The Global Fight for Climate Justice: Anticapitalist Responses to Global Warming and Environmental Destruction*. London: Resistance Books, 231–236.

Elgot, J. (2015). World Leaders Missed Chance to Tackle Climate Change, Says Economist. *The Guardian*, 25 May. Available at www.theguardian.com/books/2015/may/25/world-leaders-missed-chance-to-tackle-climate-change-says-economist.

Empson, M. (2010). British Trade Unionists Demand One Million Climate Jobs. Climate & Capitalism, 7 November. Available at http://climateandcapitalism.com/2010/11/07/british-militants-demand-one-million-climate-jobs/.

Engel-DeMauro, S. (2015). Six Questions for Ecosocialists: Response by Salvatore Engel-DeMauro. Solidarity: A Socialist, Feminist, Anti-Racist Organization, 6 May. Available at www.solidarity-us.org/ecosocialistresponseengeldemauro.

Faccer, K., Nahman, A. and Audouin, M. (2014). Interpreting the Green Economy: Emerging Discourses and their Considerations for the Global South. *Development Southern Africa*, 31(5), 642–657.

Feeley, D. (2015). Six Questions for Ecosocialists: Response by Dianne Feeley. Solidarity: A Socialist, Feminist, Anti-Racist Organization, 6 May. Available at www.solidarity-us.org/ecosocialistresponsefeeley.

Felli, R. (2014). An Alternative Socio-Ecological Strategy? International Trade Unions' Engagement with Climate Change. *Review of International Political Economy*, 21(2), 372–398.

Foran, J. (2014). Reflections on Naomi Klein's 'This Changes Everything'. Climate & Capitalism, 30 November. Available at http://climateandcapitalism.com/2014/11/30/reflections-naomi-kleins-changes-everything/.

Foran, J. (2015). *Thoughts on the Road to Paris*. Available at www.parisclimatejustice.org/sites/default/files/images/thoughtsontheroadtoparis06252015_0.pdf.

Foster, J.B. and Clark, B. (2015). Crossing the River of Fire: The Liberal Attack on Naomi Klein and 'This Changes Everything'. Climate & Capitalism, 1 February. Available at http://climateandcapitalism.com/2015/02/01/liberal-attack-on-naomi-klein-and-this-changes-everything/.

Goodman, J. and Salleh, A. (2013). The 'Green Economy': Class Hegemony and Counter-Hegemony. *Globalizations*, 10(3), 411–424.

Hahnel, R. (2012a). Desperately Seeking Left Unity on International Climate Policy. *Capitalism Nature Socialism*, 23(4), 83–99.

Hahnel, R. (2012b). Left Clouds Over Climate Change Policy. *Review of Radical Political Economics*, 44(2), 141–159.

Hahnel, R. (2013). An Open Letter to the Climate Justice Movement. New Politics, 4 November. Available from www.newpol.org/content/open-letter-climate-justice-movement.

Hamilton, C. (2017). *Defiant Earth: The Fate of Humans in the Anthropocene*. Crows Nest: Allen & Unwin.

Hornick, B. (2015). A View from Burnaby Mountain: Naomi Klein's *This Changes Everything*. Socialist Project, The Bullet, 27 March. Available from www.socialist-project.ca/bullet/1096.php.

Hornick, B. (2017). Militant Particularism and Ecosocialism: Harvey, Klein, Smith, Foster. Socialist Project, The Bullet, 26 July. Available from https://socialistproject.ca/bullet/1455.php.

Huttunen, S., Skytén, E. and Hilden, M. (2015). Emerging Policy Perspectives on Geoengineering: An International Comparison. *The Anthropocene Review*, 2(1), 14–32.

Hyman, R. (1975). *Industrial Relations: A Marxist Introduction*. London: The Macmillan Press.

International Committee of the Fourth International (ICFI) (2015). The Way Forward for Oil Workers. World Socialist Web Site, International Committee of the Fourth International (ICFI), 3 March. Available at www.wsws.org/en/articles/2015/03/03/oils-m03.html.

International Transport Workers' Federation (ITF) (n.d.). International Transport Workers' Federation homepage. Available at www.itfglobal.org/en/global/.

Johnson, R. (2015). Climate Activists Must Stand with Striking U.S. Oilworkers. Climate & Capitalism, 12 February. Available at http://climateandcapitalism.com/2015/02/12/climate-activists-must-stand-striking-u-s-oilworkwers/.

Kahle, T. (2014). Rank-and-File Environmentalism. *Jacobin*, 11 June. Available at www.jacobinmag.com/2014/06/rank-and-file-environmentalism/.

Kahle, T. (2015). The Seeds of an Alternative. *Jacobin*, 19 February. Available from www.jacobinmag.com/2015/02/refinery-workers-strike-ecosocialism/.

Klein, N. (2014). *This Changes Everything: Capitalism vs The Planet.* London: Allen Lane.

Kovel, J. (2007). *The Enemy of Nature: The End of Capitalism or the End of the World?* New York: Zed Books.

Light, J. (2015). Greens Get Behind Striking Oil Workers. Grist, 3 February. Available at https://grist.org/climate-energy/greens-get-behind-striking-oil-workers/.

Lipow, G. (2014). Zombie Carbon Trading's Latest Resurrection. Grist, 15 January. Available at http://grist.org/article/zombie-carbon-tradings-latest-resurrection.

Lipow, G. (2015). Shutting Down the Fog Machine. *Review of Radical Political Economics*, 47(2), 231–242.

Löwy, M. (2015). *Ecosocialism: A Radical Alternative to Capitalist Catastrophe.* Chicago, IL: Haymarket Books.

Lusanne, J. (2015). Limited Strike Called at US Oil Refineries as National Contract Expires. World Socialist Web Site, International Committee of the Fourth International (ICFI), 2 February. Available at www.wsws.org/en/articles/2015/02/02/usw-f02.html.

Luscombe, R., Pilkington, E. and Smith, D. (2017). Florida Officials Warn Irma Will be 'Storm Wider than the State'. *The Guardian*, 8 September. Available at www.theguardian.com/world/2017/sep/08/florida-officials-warn-irma-will-be-storm-wider-than-the-state.

McDonald, M. (2015). The 'Climate Games' Aren't Just Activist Stunts – They're Politics Beyond the UN. The Conversation, 9 December. Available at https://the-conversation.com/the-climate-games-arent-just-activist-stunts-theyre-politics-beyond-the-un-51872.

McIlroy, J. (2014). Marxism and the Trade Unions: The Bureaucracy versus the Rank and File Debate Revisited. *Critique: Journal of Socialist Theory*, 42(4), 497–526.

McIntyre, R. and Hillard, M. (2012). Capitalist Class Agency and the New Deal Order: Against the Notion of a Limited Capital-Labor Accord. *Review of Radical Political* Economics, 45(2), 129–148.

Pilkington, E. (2017). A Tale of Two Irmas: Rich Miami Ready for Tumult as Poor Miami Waits and Hopes. *The Guardian*, 9 September. Available at www.theguardian.com/world/2017/sep/09/hurricane-irma-miami-florida-two-cities.

Riddell, J. (2014). Naomi Klein: 'Only Mass Social Movements Can Save Us'. Climate & Capitalism, 19 October. Available from http://climateandcapitalism.com/2014/10/19/naomi-klein-climate-change-mass-social-movements-can-save-us/.

Rugh, P. (2015). Striking for Climate Justice. *Dissent Magazine*, 21 February. Available from www.dissentmagazine.org/blog/usw-oil-workers-strike-climate-justice-solidarity.

Smith, J. and Wiest, D. (2012). *Social Movements in the World-System: The Politics of Crisis and Transformation.* New York: Russell Sage Foundation, New York.

Smith, R. (2014). Climate Crisis, the Deindustrialization Imperative and the Jobs vs. Environment Dilemma. Truthout, 12 November. Available from www.truth-out.org/news/item/27226-climate-crisis-the-deindustrialization-imperative-and-the-jobs-vs-environment-dilemma.

Smith, R. and Foster, J.B. (2017). The Significance of Naomi Klein: An Ecosocialist Exchange. Climate & Capitalism, 4 May. Available from http://climateandcapitalism.com/2017/05/04/significance-naomi-klein-ecosocialist-exchange/.

Solidarity (2015). Six Questions for Ecosocialists: Introduction. Solidarity: A Socialist, Feminist, Anti-Racist Organization, 6 May. Available at https://solidarity-us.org/ecosocialistquestions/.

System Change Not Climate Change (SCNCC) (n.d.). *What is Ecosocialism?* System Change Not Climate Change. Available at http://systemchangenotclimatechange.org/sites/default/files/images/scncc-pamphlet-v2.pdf.

System Change Not Climate Change (SCNCC). (2015a). Climate Space Events in Paris. System Change Not Climate Change, December. Available at https://climatespace2013.files.wordpress.com/2015/12/cs-events-flyer3.pdf.

System Change Not Climate Change (SCNCC). (2015b). Paris Climate Justice: Climate Justice News, Analysis and Action. System Change Not Climate Change, December. Available at www.parisclimatejustice.org.

System Change Not Climate Change (SCNCC) (2015c). *Redlines Are Not for Crossing: A Global Blockadia Proposal for the D12 Day of Action, at the End of the Paris COP21 Summit.* System Change Not Climate Change. Available at www.parisclimatejustice.org/sites/default/files/images/redlinesproposalsept.pdf.

System Change Not Climate Change (SCNCC) (2017a). Seven Questions for Ecosocialists. System Change Not Climate Change. Available at http://systemchangenotclimatechange.org/seven-questions-ecosocialists.

System Change Not Climate Change (SCNCC) (2017b). System Change Not Climate Change: An Ecosocialist Coalition homepage. Available at http://systemchangenotclimatechange.org/.

Temper, L. and Gilbertson, T. (eds) (2015). *Refocusing Resistance to Climate Justice: COPing In, COPing Out and Beyond Paris.* EJOLT Report, 23. Available at www.ejolt.org/2015/09/refocusing-resistance-climate-justice-coping-coping-beyond-paris/.

Trade Unions for Energy Democracy (TUED) (n.d.). Trade Unions for Energy Democracy Homepage. Available from http://unionsforenergydemocracy.org/.

Turnbull, D. (2015). Solidarity with Striking Refinery Workers. Oil Change International, 2 February. Available at http://priceofoil.org/2015/02/02/solidarity-with-refinery-workers/.

Uehlein, J. (2015). Labor Network for Sustainability Calls for Support for Oil Strikers. Available at www.labor4sustainability.org/articles/labor-network-for-sustainability-calls-for-support-for-oil-strikers/.

Weedall, G. (2013). Why We Need an Ecosocialist Revolution. *Green Left Weekly,* 1 February. Available at www.greenleft.org.au/content/why-we-need-ecosocialist-revolution-0.

Weston, D. (2014). *The Political Economy of Global Warming: The Terminal Crisis.* New York, Routledge.

White, J. (2015). Oil Strike Shows Growing Combativeness of US Workers. World Socialist Web Site, International Committee of the Fourth International (ICFI), 9 January. Available at www.wsws.org/en/articles/2015/02/09/stri-f09.html.

Williams, C. (2013). Strategy and Tactics in the Environmental Movement. Climate & Capitalism, 21 September. Available at https://climateandcapitalism.com/2013/09/21/strategy-tactics-environmental-movement/.

Zill, Z. and Ware, M. (2014). A New Stage for Ecosocialist Unity and Action. Climate & Capitalism, 2 February. Available at https://climateandcapitalism.com/2014/02/02/new-stage-ecosocialist-unity-action/.

7 The biosphere and social forces in a geopolitically unstable world beset by organic crisis

Introduction

This chapter draws on relevant ecosocialist writings and on academic literature from a range of discipline areas to locate ecosocialist struggles for climate justice within the wider context of dominant social forces in the current world order. The ecological neo-Gramscian analytical framework introduced in Chapter 2 is applied under sub-headings that reflect a dialectical analysis which shifts the focus of the discussion from how social systems affect the biosphere to how the changed biosphere affects social systems. Reversing the dominant view of mainstream economists that 'nature' is a passive element and a 'subset' of the economy (which is also a normative perspective adopted by political leaders, policymakers and many academics – including IR and IPE theorists), in the 'MHS Forces Redux II' framework the capitalist mode of production dominating the global economy is understood as necessarily existing within the Earth's biosphere (or 'nature'). The global economy, like all social systems, is thus recognised as being subject to the restrictions of the objective laws of physics that living within a natural biosphere impose.

The relationship between the Earth's biosphere and human production and reproduction is, moreover, part of a single complex and evolving dynamic system; while the productive and reproductive capabilities of all life-forms are embedded within the confines of the biosphere, in a global economy where the capitalist mode of production dominates, a relatively small proportion of humans benefit from productive activities that radically change this biosphere, and the transformed biosphere in turn affects the productive and reproductive capabilities of all life-forms (including the vast majority of humans who are not responsible for the damage to the biosphere). The analysis that follows emphasises the ways in which changes in the biosphere impact on the material reproductive and productive capabilities of the humans and social systems embedded in it, thus providing a case study demonstrating the utility of the ecological neo-Gramscian theoretical perspective at the world order level.

The current world order: a non-hegemonic world order in organic crisis

Emphasising its consensual aspects, Cox (1987) uses Gramsci's definition of hegemony to categorise different structures of world order in the past one-and-a-half centuries as hegemonic and non-hegemonic.[1] In his 1987 monograph, Cox identifies the 1789–1873 liberal world order and the post–World War II capitalist world orders as hegemonic, and the 'era of rival imperialisms' (1873–1945) as non-hegemonic because of the instability that resulted in the twentieth century's two world wars. Writing in 1987, Cox refers to the first hegemonic world order as *Pax Britannica* because Britain reigned supreme and shaped the global economy in its interests; the second hegemonic world order, *Pax Americana*, began when the United States emerged as the clear victor of World War II. This categorisation replaced Cox's initial tentative (1981, 1983) proposal that 1965 signalled the relative decline of US power, and hence the end *of Pax Americana*. Commenting on developments that followed these earlier writings, Cox adopted the view that US hegemony in world affairs still prevailed into the early twenty-first century, after the modifications to the global political economy that began in the mid-1970s (Cox with Schechter, 2002). In the 1990s, Cox (1996, p. 155) identified a number of issues that capitalist globalisation could exacerbate, including migration, social polarisation and ecological issues such as pollution and the over-exploitation of non-renewable resources. Since then, many of these issues have indeed combined to become trigger-points for social conflict and there is over-whelming evidence that the current capitalist world order is experiencing an organic (permanent and system-wide) crisis as a result of the serious and widespread ecological, economic and socio-political harms it engenders as it continues its 'normal' operations.

In interviews conducted after the 2008 Global Financial Crisis (GFC), Cox argued that the US-led neoliberal globalisation project no longer enjoyed widespread support, as evidenced in the popular protests that occurred both in the US and in European countries such as Italy and Greece in the GFC's aftermath (Martin, 2013). This is because, rather than represent-ing a 'universal general interest' whereby a majority of classes and social groups benefit, capitalism overwhelmingly favours a global elite comprising of a nascent 'transnational capitalist class' (TCC) and an allied 'transnational managerial class' (TMC) at the expense of the planet and of the wellbeing of the vast majority of people. In short, the current world order is categorised as non-hegemonic because the project of capitalist globalisation that prevails cannot make the concessions that would elicit the widespread support from global civil society that, according to Cox and other neo-Gramscian analysts, is the prerequisite of hegemony. At the global level, Cox argued that there is also much disquiet about the ability of the US to continue its role as global hegemon, with tensions between the US and an economically ascendant China, and between the US and Russia, being indicators of the United

States' declining power (Schouten, 2009). These observations led Cox (1996, p. 155) to conclude that 'globalization is not the end of history but the initiation of a new era of conflicts and reconciliations'.[2] Echoing Naomi Klein's observation about the unfortunate coincidence of the scientific discoveries of anthropogenic global warming (AGW) in the same decade as the extension of the neoliberal globalisation project, it could be argued that this new era of conflicts could not have come at a worse time in the history of humanity – at a time when addressing the challenges of the climate crisis and the general degradation of the biosphere in the Anthropocene requires an unprecedented level of cooperation between nations and peoples.

Material capabilities: the capitalist mode of production and the Anthropocene's unstable biosphere

As noted in previous chapters, the impacts of human activities on the biosphere are now so widespread and persistent that they have prompted Earth System scientists to propose the formal demarcation of a new geological epoch: the Anthropocene. Ecosocialists have for many years argued that the capitalist mode of production has particularly damaging effects on the environment because, amongst other reasons, capitalism is motivated by competition and profit maximisation, which entail minimising costs and focusing on short-term financial gains. This exploitation of both people and nature spreads geographically as capital expands its operations into 'underdeveloped' regions of the world.[3] The global expansion of capitalism occurs unevenly and follows geographically and historically specific trajectories that differ across space and time because 'once capitalism is established in one part of the world it affects and changes the form of transition to capitalist development elsewhere' (Ashman, 2009, p. 36). In two seminal works, the *History of the Russian Revolution* and *The Permanent Revolution*, Leon Trotsky expands on and deploys the concept of 'uneven and combined development' as an analytical tool for understanding these characteristics of capitalism.[4]

Applying the concept of 'uneven and combined development' to the environmental issues that are the focus of his analysis, ecosocialist James O'Connor discusses how these features of the global expansion of capitalism not only spread but also intensify the environmental degradation inherent in capitalist relations of production. Emphasising unequal power relationships, O'Connor defines 'uneven development' with reference to development characterised by exploitative relationships 'between town and country (centre/periphery; developed/underdeveloped country)' and 'combined development' with reference to how the geographical growth of capitalist relations of production combine 'the most profitable features of development [such as 'advanced technology, industrial organisation and division of labour'] and underdevelopment [such as a disciplined and cheap labour force] in a new unity which maximises profit increases' (O'Connor, 1989, pp. 1–2).

Some of the environmentally damaging effects of uneven development that O'Connor identifies in rural, periphery and underdeveloped regions include: deforestation (primarily as a result of clearing land to facilitate agribusiness) and the associated environmental problems this causes (for example, soil loss, aridity, and droughts); the intense exploitation of resources such as fossil fuels and other minerals and the pollution associated with mining and processing these resources; and the degradation of land and waterways associated with agribusiness practices of monoculture and the usage of polluting chemical fertilizers and pesticides. 'Combined development', whereby capital relocates manufacturing operations to underdeveloped regions with low wages and lax environmental regulations, damages previously 'underdeveloped' Global South environments by exporting both dangerous products (such as chemicals banned in the advanced capitalist countries) and pollution from the Global North. O'Connor concludes that:

> When uneven and combined development of capital are themselves combined, it would appear that super-pollution in industrial zones may be explained by super-ecodestruction of land and resources in raw material zones, and vice versa. Depletion and exhaustion of resources and pollution depend on one another; they are the necessary result of the same universal process of capital 'valorisation'. Depletion/exhaustion and pollution are thus not independent issues. The natural wealth of the world is depleted and turned into garbage, often dangerous garbage, through global capital accumulation. And the unwanted by-products – pollution – have the effect of depleting/exhausting resources. Put formally, the greater the profit rate, the greater the accumulation rate, the greater the rate of depletion/exhaustion which indirectly leads to a greater rate of pollution.
>
> (O'Connor, 1989, pp. 10–11)

While the environmental degradation and destruction that are a natural part of capitalist relations of production take a variety of forms that affect all ecosystems, ecosocialists have been receptive to, and fully support, the arguments of many scientists that global warming, climate change and ocean acidification are the most pressing symptoms of the anthropogenic damage to the biosphere in the current conjuncture and for the foreseeable future. The primary cause of global warming and ocean acidification is the widespread use of fossil fuels that emit CO_2, the predominant GHG currently forcing global warming (IPCC, 2014).[5] Fossil fuels are essential in the operations of global capitalism as they are used not only for generating the energy required for the production of goods and services but also for the transportation of goods and people around the globe (Tanuro, 2013). In addition, fossil fuels and products derived from them are essential in the operations of large-scale agribusiness practices (Magdoff, 2015), which also contribute substantially to the emission of two other extremely potent and long-lasting GHGs, methane and nitrous oxide (IPCC, 2019a).

Apart from raising the average global temperature, the energy imbalances that result from a greater concentration of GHGs in the atmosphere have a variety of effects on the complex and interconnected Earth System (Hansen et al., 2016; Harrison et al., 2016).[6] An example of how changes due to rising temperatures in one part of the Earth System affect other parts is evident if one considers the way in which the melting of ice sheets not only results in sea level rise, but also increases the volume of low-density freshwater in the oceans (Hansen et al., 2016). This change in the composition of the water in the oceans affects global ocean circulation patterns (Hansen et al., 2016; Liu et al., 2017), which in turn affect cloud formation (Norris et al., 2016) and thus precipitation patterns (Chadwick et al., 2016). In addition, as the oceans absorb CO_2, ocean acidification has increased to unprecedented levels (IPCC, 2019b). Moreover, ocean acidification exacerbates other problems arising from large-scale capitalist economic activity, such as the commercial depletion of fish stocks and the disruption of the nitrogen cycle as a result of large-scale, fertilizer-intensive commercial agricultural practices (UNEP, 2016). In summary, research unequivocally demonstrates that anthropogenic GHG emissions and the resultant global warming, in addition to many other environmentally damaging economic practices, are radically changing ecosystems, oceans and the biosphere of which these subsystems are a part. These changes have consequences (many of them adverse) for all life-forms that have evolved to live in the relatively stable climatic conditions of the pre-Anthropocene biosphere that existed during the Holocene (Williams and Crutzen, 2013).

Neoliberal capitalist institutional responses to the need to reduce GHG emissions and maintain a habitable planet

Despite even neoliberal institutions such as the World Bank (2014) and the World Economic Forum (WEF, 2020) identifying global warming as a high-ranking issue of concern amongst their lists of 'threats' to the global economy, the neoliberal agenda of further developing a global political economy with minimal restrictions on capital accumulation continues to encourage practices that cause serious environmental damage both directly and indirectly. Examples of such practices include the building of new fossil fuel power plants (Shearer et al., 2016), the implementation of environmentally damaging trade treaties (Aykut, 2016; Gallagher, 2016) and, more generally, the relentless promotion of a narrowly conceived 'economic growth' ideology that relies on ever-increasing consumerism (Foster, Clark and York, 2010). Scientists such as Rockström et al. (2016, p. 469) argue that 'each day without a zero carbon roadmap increases the stakes in our global climate gamble', and leading climate scientist and prominent climate movement activist James Hansen has gone as far as calling the Paris Agreement a 'fake' and a 'fraud' (Milman, 2015).

Material capabilities: the biosphere's responses to capitalist profit-maximising practices

The physical responses of the actually existing Earth System to 'business as usual' conform entirely to the laws of physics (Nuccitelli, 2017) and include: ice sheets melting and causing sea levels to rise; ocean acidification and warmer oceans bleaching coral reefs; and 'extreme weather' events in the form of excessive precipitation and flooding, extreme and prolonged heatwaves, prolonged droughts and the increasing frequency of widespread and poorly understood 'extreme wildfires' (Arctic Council, 2016; IPCC, 2019b; Sharples, 2016; WMO, 2019). These physical manifestations of the Earth's changing biosphere devastate natural ecosystems and communities that live in them and rely on them for their livelihoods and survival (WHO, 2016). Given the logic of the 'non-negotiability' of the laws of physics, how is one to understand the failure of authorities to take decisive and effective action? Ecosocialists argue that beyond the power of the fossil fuel industry, wider issues of capitalist social relations of production and of class dynamics are the underlying reasons for the inaction of policymakers despite the fact that it is in the realm of human capabilities to address one of the greatest threats humanity has ever faced as a species (Angus, 2016; Weston, 2014).

Material capabilities: class and reproductive capabilities in the Anthropocene

As ecosocialist Ian Angus (2016) bluntly points out, with respect to the effects of global warming and climate change, 'we are not all in this together'.[7] In the short term at least, and perhaps even in the medium term, affluent people with sufficient resources can insulate and protect themselves from the worst immediate effects of global warming and climate change (Angus, 2016; see also di Muzio, 2015). However, given the uncertainties involved, it is highly likely that even the wealthiest and most privileged will not be able to 'buy' their way out of the longer-term effects of a changing biosphere if it shifts to a different state. Scientists warn that changes within complex systems are non-linear and unpredictable, and that sudden shifts within the physical Earth System could take everyone by surprise if poorly understood planetary boundaries are crossed.[8] A graphical depiction of some potential Earth System 'tipping elements' and some of the possible interactions between them are shown in Figure 7.1.

Despite the currently dominant factions of global elites and policymakers choosing to limit their responses to these unfolding ecological disasters to woefully inadequate, voluntary and incremental measures that are not legally binding and that experts calculate will result in rising average temperatures of 3.2°C even if fully implemented (UNEP, 2017), Naomi Klein is correct to conclude that when it comes to climate change, 'this changes everything' (Klein, 2014); no amount of politicking can change the reality of how the

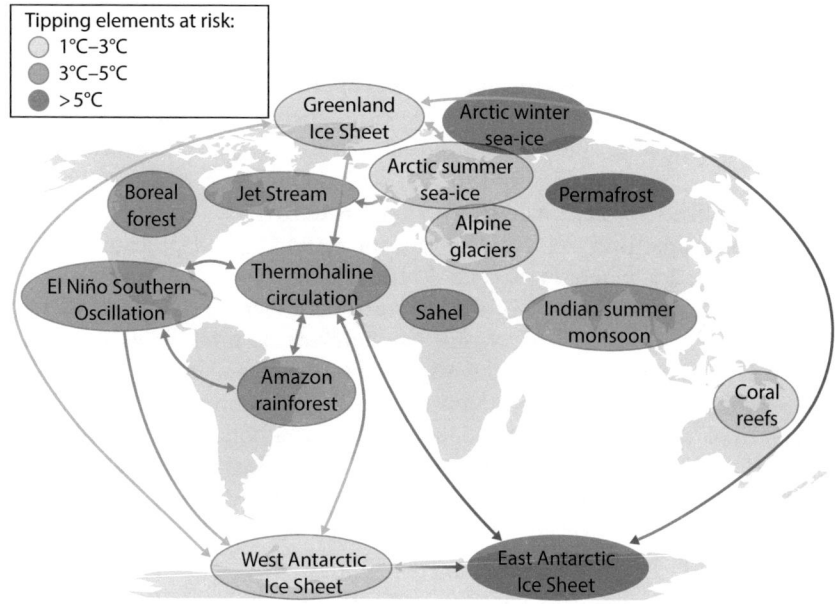

Figure 7.1 Earth System tipping elements at risk.

actually existing Earth System responds to physical inputs and outputs. As Steffen et al. (2011) point out when comparing the ideologically informed rejection of Darwinian evolution to a similarly ideologically based denial of the urgency of reducing anthropogenic GHG emissions in order to mitigate climate change:

> Darwin's insights into our origins provoked outrage, anger and disbelief but did not threaten the material existence of society of the time. The ultimate drivers of the Anthropocene, on the other hand, if they continue unabated through this century, may well threaten the viability of contemporary civilization and perhaps even the future existence of *Homo sapiens.*
>
> (Steffen et al., 2011, p. 862)

While many of those with the wealth and power to control world affairs apparently feel they are exempt from the laws of physics and believe themselves to be invincible (perhaps because they rely on some future, as-yet-undiscovered or undeveloped technological innovation to save the day), ecosocialists and climate justice movement actors are concerned about how many of the least powerful, poorest and most vulnerable people who do not even contribute to global warming are the first to suffer from its effects.

Damaging effects of AGW-induced climate change range from changing weather patterns disrupting normal agricultural production and threatening food supplies (WMO, 2019) to increasingly powerful typhoons killing thousands and leave millions homeless (Campbell, 2013). It is not only disadvantaged people in the Global South that suffer the consequences of 'business as usual'; poor and vulnerable people and communities in the so-called 'advanced' capitalist countries, who can ill afford to deal with additional crises, also face the devastating effects of the changing weather patterns and 'extreme weather events' that result from AGW. The lack of resources and preparedness of poor communities when Hurricane Katrina struck New Orleans in 2005 greatly exacerbated the effects of the disaster (Schlosberg and Collins, 2014), and poorer people also suffered worse consequences as a result of Hurricane Sandy seven years later. In the aftermath of Hurricane Sandy, the realities of capitalist social relations asserted themselves in how 'reconstruction assistance was allocated disproportionately to homeowners rather than tenants, even though the latter were more likely to be in the lower-income bracket' (IDMC, 2015, p. 51), and how over 39,000 people who had to evacuate their homes in 2012 because of what came to be known as 'Superstorm Sandy' were still in need of housing assistance in 2015 (*ibid.*, p. 93). These examples illustrate that whether they live in the Global North or in the Global South, the already precarious existence of disadvantaged and poor communities further exacerbates their vulnerability to the effects of AGW and the resultant 'extreme weather' (Leichenko and Silva, 2014).

Given the evidence, ecosocialists argue that climate change is not a Global North/Global South issue; it is a *class* issue because in a capitalist system the material resources that individuals and communities have access to determine their life chances in a variety of ways, including in how effectively they can cope with the ravages of 'extreme' weather events. These 'extreme' weather events are, moreover, occurring at a time when the reorganisation of the material forces of production and the shifting balance of power in the social forces of production (that is, in the balance of power between capital and labour) result in more and more Global North workers joining the ranks of the disadvantaged communities of the Global South, 'falling' into precarious existences and poverty as governments continue to implement 'flexible labour market' and wage repression policies that lead to underemployment and the casualisation of work (Heyes, Lewis and Clark, 2012).

The balance of social forces and its implications for the reproductive capabilities of labour and subaltern groups in a global capitalist world order

The uneven and combined development that characterises the geographical expansion and intensification of capitalist social relations of production has important outcomes vis-à-vis the relative strength of global capital and global labour. The labour movement in the advanced capitalist economies is now relatively weaker than it used to be in the 1960s and 1970s because of the greatly enlarged global reserve army of labour, which is a consequence not only of the increasing proletarianisation of populations in the Global South (Bieler, 2012) but also of the incorporation of China and the previous Eastern Bloc countries into the global capitalist economy (Jefferies, 2015).[9] Technological and managerial innovations facilitating changes in the organisation of how and where goods are produced, together with policies implemented by national governments to create 'flexible labour markets' and facilitate the free flow of capital across borders, shift the labour force to countries where organised labour is weak while simultaneously greatly compromising the ability of the working class in the advanced capitalist economies to defend their jobs, wages and working conditions (Harvey, 2010). The role of institutions and social facts in the implementation of these changes to the operations of the global economy should not be underestimated. As many theorists point out, the changes are, to a large extent, the result of a globally conceived and implemented neoliberalising project which takes specific forms in different places and at different times, depending on the historically shaped 'local' national conditions.

Despite spatial and temporal variations in the expansion and intensification of global capital, the United Nations Conference on Trade and Development (UNCTAD) 2016 Trade and Development Report states that there has been a general global increase in inequality of income distribution since the late 1970s and early 1980s. According to the same report, this growing inequality has been accompanied by a rising trend of profit-seeking by depressing wages and financial rent-seeking rather than an increase in industrial activity through innovation and investment. In addition, privatisation, deregulation and lower public expenditures on essential social services such as healthcare and education further impoverish wage earners (especially the most vulnerable, who depend on state programmes), while simultaneously increasing profit opportunities in the private sector by providing new areas for capital investment (UNCTAD, 2016). While UNCTAD's focus is primarily on 'developing' countries, the report also refers to 'stagnant demand' for the commodities produced in the Global South due to austerity policies implemented in some countries after the 2007/2008 GFC and rising unemployment and precarious employment in the Global North (UNCTAD, 2016), trends confirmed by several other analysts (for example, Cronin, 2013; Piketty, 2014).

Drawing on evidence gleaned from a variety of official statistical data sources, economist Thomas Piketty (2014, p. 471) points out that 'in the second decade of the twentieth century, inequalities of wealth that had supposedly disappeared [in the twentieth century after, and largely due to, the two world wars] are close to regaining or even surpassing their historical highs'. More importantly, his historical investigation reveals that inequality is an inherent feature of capitalism because 'wealth accumulated in the past grows more rapidly than output and wages' (Piketty, 2014, p. 571). In short,

Piketty points out that an initially unequal distribution of wealth results in increasing wealth inequalities, both logically and in practice.[10] A January 2017 briefing paper published by Oxfam makes the same point, but more dramatically; it claims that only eight men own as much wealth as the poorest 50% of the global population (Oxfam, 2017). Another staggering statistic in the same Oxfam briefing paper is that the richest 1% has owned more wealth than the rest of the global population (the 99%) since 2015. Evaluating the integrity of these claims, Australian National University Professor Peter Whiteford concludes that they are supported by the best available data and that, even if the disparities between the wealthiest minority and the poorest majority are not as great as reported in the briefing paper, they are nevertheless 'massive' (Whiteford, 2017).

Rising inequality is a concern to some mainstream economists (the organic intellectuals of the ruling class, in Gramscian terminology) because it can contribute to economic crises, and hence also to political crises and social instability that could threaten global capitalism as a system. For example, Piketty argues that rising inequality was a contributing factor leading to the 2007/2008 GFC. Economists such as Piketty are well aware of the political dangers of this situation, as is evident in the following statement.

> Taxation is not a technical matter. It is pre-eminently a political and philosophical issue, perhaps the most important of all political issues. Without taxes, society has no common destiny, and collective action is impossible. This has always been true. At the heart of every major political upheaval lies a fiscal revolution. The *Ancien Régime* was swept away when the revolutionary assemblies voted to abolish the fiscal privileges of the nobility and clergy and establish a modern system of universal taxation. The American Revolution was born when subjects of the British colonies decided to take their destiny in hand and set their own taxes. ('No taxation without representation'.) Two centuries later the context is different, but the heart of the issue remains the same.
>
> (Piketty, 2014, pp. 492–493)

Using a problem-solving approach, Piketty thus argues for the implementation of reformist policies such as progressive taxation and a global tax on capital because of his keen awareness of the role of political democracy as a justification for capitalism – for example, he states the following as an 'essential truth': 'defining the meaning of inequality and justifying the position of the winners is a matter of vital importance, and one can expect to see all sorts of misrepresentations of the facts in service of the cause' (Piketty, 2014, p. 487).

Piketty points out that 'in a democracy, the professed equality of rights of all citizens contrasts sharply with the very real inequality of living conditions, and in order to overcome this contradiction it is vital to make sure that social inequalities derive from rational and universal principles rather than arbitrary contingencies. Inequalities must therefore be just and useful to all, *at least in the realm of discourse and as far as possible in reality as well'* (Piketty, 2014, p. 422; emphasis added). The role of the organic intellectual of the ruling class as justifier of class inequalities is transparent in this particular statement; even if social inequalities themselves cannot be avoided, they *must be seen as being 'just and useful to all'*, even if only 'in the realm of discourse'! This exhortation perhaps constitutes an implicit acknowledgement by Piketty that the structural 'problems' he refers to in his monograph are deeply embedded within the global capitalist economy's operations and may prove impossible to reform given existing power relationships. As Heyes, Lewis and Clark (2012) argue, governments now depend on 'finance-led growth' both directly (because of the jobs and tax revenues the finance sector generates) and, perhaps more importantly, indirectly; finance capital provides 'cheap, unsecured credit' that boosts effective demand in economies that would otherwise stagnate because of the low wages imposed on workers.

Forms of state and social forces in a neoliberalising capitalist world order

The reasons for governments' dependence on finance-led growth are deeply structural and can be explained with reference to the financialisation of capitalist economies since the early 1980s (Lapavitsas and Mendieta-Muñoz, 2016). Lapavitsas (2013) argues that because large corporations frequently use retained profits to finance their investments and also engage in independent financial operations and trading, the finance sector has positioned itself to find alternative sources of profits: transactions in financial markets, fees and commissions it charges as intermediary in financial transactions and transactions with households and individuals. Financialised capitalism relies on the deep penetration of finance into household and individual revenues that enable what Lapavitsas (2014, p. 37) refers to as the 'financial expropriation' of profits 'directly from wages and salaries' in the form of interest made on household mortgages and on individuals' unsecured consumer loans. Not only are individuals forced to borrow to meet basic needs in order to supplement low incomes and because of the state's retreat from public provisioning of health, education and other essential services, but they are also forced to hold 'substantial financial assets' themselves because of the privatisation of previously public pension funds (Lapavitsas, 2013).

By implementing ideologically informed neoliberal financial and labour market deregulation policies, and by restructuring tax regimes in favour of the rich and thus eroding the revenue base with which social provisioning can be funded, national governments have been complicit in the central position that finance capital now enjoys in national economies (Bruff, 2014; Keaney, 2014; Lapavitsas, 2013, 2014). States also back powerful independent central banks that set 'benchmark interest rates' and use public money to provide liquidity to the finance industry in times of crisis, thus giving 'a vast public subsidy to the capitalist class as a whole' (Lapavitsas and Mendieta-Muñoz, 2016, p. 51). These features of capitalism imply that despite concerns about the political fallout of rising inequality raised by economists such as Piketty, who try to defend liberal democratic ideals in their attempts to save the capitalist system from itself, it is not clear how the system can be 'reformed' given the power of corporate and financial interests and the way in which these interests successfully counter any attempts to regulate or control their operations – attempts that government officials and policy-makers in any case make only half-heartedly (Helleiner, 2013; Keaney, 2014; Peck, 2013).

Despite the historical evidence he has amassed and discusses at length in his book, Piketty remains 'optimistic' that democracy can 'regain control over capitalism', although he admits that there is no basis for this optimism.

> Has the US political process been captured by the 1 percent? This idea has become increasingly popular among observers of the Washington political scene. For reasons of natural optimism as well as professional predilection, I am inclined to grant more influence to ideas and intellectual debate.
>
> (Piketty, 2014, p. 513)

The evidence that Piketty himself refers to in *Capital in the Twenty-First Century* indicates that 'representative democracy' is little more than a hollow concept since it represents the interests of wealthy elites and the well-paid and powerful managers that manage their affairs rather than the general interest. Moreover, there is much evidence that people suffering the effects of policies that favour 'the 1%' are aware of their situation and no longer trust mainstream politicians and the major political parties to represent their interests (Keaney, 2014; Peck, 2013); the economic crisis has become a political crisis of legitimacy, which provides further evidence that the current crisis is organic (in the Gramscian sense) rather than conjunctural and presents a system-wide weakness that erodes the consensus on which hegemony depends.

Forms of state, institutions and social facts: liberal representative democracy as a 'hollow concept' in a neoliberalising capitalist world order

With governments implementing policies that have, over the years, favoured large and powerful financial institutions and multinational corporations, national economies have become increasingly intermeshed and dependent on these firms and institutions. As a result, government representatives and policymakers declare themselves unable to secure their citizens' social and economic wellbeing except in the limited way of accommodating capital in order to attract investment to 'strengthen' the economy and thereby to secure jobs (Bruff, 2014; Carroll and Jarvis, 2015), and for many years large segments of the populations of Global North societies seemed to have accepted this 'truth' that 'there is no alternative' so that it became what Sinclair refers to as a 'social fact'. But social facts are not immutable, and the 2007/2008 GFC initiated a widespread questioning of the expansion and intensification of global capitalism (Brands and Feaver, 2016). As the economic and social effects of the financial crisis that originated within US financial institutions worked their way throughout the global economy (Karanikolos et al., 2013), the unwillingness of governments (with the notable exception of Iceland) to discipline the financial sector became particularly evident.[11] To many people, governments' unilateral decisions to use taxpayer money and to put taxpayers into further debt in order to bail out the rich, when the same governments have been arguing for decades (and continue to argue) that they do not have the money to support essential social programmes that many working-class and poor people rely on just to get by, demonstrated that the major political parties are themselves responsible for implementing and defending policies that favour the rich at the expense of everyone else (Keane, 2013; Ortiz et al., 2013).

Academic research provides evidence supporting the widely recognised popular understanding that economically powerful elites and organised business groups strongly influence US government policy 'while average citizens and mass-based interest groups have little or no independent influence' (Gilens and Page, 2014, p. 564). This systemic deficit in political representation for 'average citizens' is not confined to the United States; it is a widespread phenomenon characterising many Western democracies (Chou, 2015; Matthijs, 2014). One response by ordinary people has increasingly manifested as a tendency to use the political institutions at their disposal to vote against global elites whenever they get an opportunity to do so, with the 'Brexit' outcome of the June 2016 UK referendum on European Union membership and the November 2016 election of populist Republican maverick Donald Trump as US president being examples (Desai, Freeman and Kagarlitsky, 2016; Waddock, 2016).

Populist politicians such as Trump are elected on the basis of their nationalist and anti-immigrant rhetoric, jingoism and promises that they will create local jobs or implement other policies to protect national populations against the ravages of neoliberal globalisation (Fouskas, 2016), but it is unlikely that they will meet their election promises once in power (Widmaier, 2016).[12] An illustrative example of the powerlessness of voters in liberal democratic political systems to effect systemic change through formal democratic institutions is evident in the ongoing drama of the continuing economic crisis in Greece. While the majority of the Greek people clearly voted Syriza – which grew out of the Radical Left Coalition formed in 2004 and emerged as an electoral political party in 2012 – into power with the mandate to end austerity policies (Witte, 2015), the Syriza government proved unwilling or unable to follow the people's mandate and continued, instead, to impose the increasingly severe and socially damaging austerity policies dictated by what Greek people disparagingly call 'the Troika' (the International Monetary Fund, the

European Union and the European Central Bank). The inability or unwill-ingness of official political representatives to address the economic, environ-mental and social damage caused by the expansion and intensification of global capitalist relations of production has forced many to turn to alternative means of expressing their frustration. As Robert Cox put it, 'people don't believe in politics any more' (Cox cited in Martin, 2013, p. 221).

The social dynamics of protest in a non-hegemonic neoliberalising capitalist world order

Unable to achieve meaningful change through legal political institutions, an increasing number of people resorted to participating in civil disobedience mass actions such as the 2011/2012 Occupy Movement protests (della Porta, 2012). While the relatively long-lasting Occupy camps in US and UK public squares and parks were forcibly disbanded in what appeared, at least in the US, to be a coordinated action by authorities (Ramsey, 2012), protests over a variety of eco-nomic, environmental and social issues continue to occur in many parts of the world (Caraus and Parvu, 2017; Youngs, 2017). The early days of the Trump Administration in particular heralded a renewal of mass-protest activities both in the United States and worldwide (Jamieson, 2017). Notable mass-protest actions against President Trump and his administration's policies included the 15 April 2017 Tax March demanding that he release his tax returns (Stevens, 2017), the 22 April 2017 March for Science to counter the Trump Administration's repeal of environmental protection laws and its funding cuts to environmental protec-tion agencies and research projects on crucial issues such as climate change (Milman, 2017), the 29 April 2017 People's Climate March demanding socially just action on climate change (Fandos, 2017) and the 12 August 2017 Charlottesville (Virginia) protest against the rise of US right-wing extremism, which is perceived as being fuelled by the Trump Administration's policies (Alpher, 2017). At the transnational level, an estimated 200,000 anti-capitalist protesters gathered in force in Hamburg on 7 and 8 July 2017 to protest against the G20 Summit (Oltermann, 2017) in an action reminiscent of the 1999 alter-globalisation anti-WTO protests in Seattle (Price, 2016).

Ongoing sporadic protests such as these demonstrate that an increasing number of people belonging to, or supporting, subordinate groups that include large segments of the working class, refugees, migrants, minority groups, students and people with disabilities are becoming concerned enough to take action in the form of protests against the negative material and social outcomes that they have to suffer as a result of policies that favour elites (CIVICUS, 2016). Few of the protests have resulted in changes that benefit ordinary people; on the contrary, ruling elites respond with thinly veiled disdain for the concerns of their citizens, as they did in the aftermath of the 2003 global anti-war demonstrations, in which millions of people participated both in the US and in many cities around the world (Carty, 2009; Hil,

2008).[13] These global anti-war demonstrations signalled people's opposition to the 2003 military invasion of Iraq by the 'Coalition of the Willing' – an invasion led by the US Bush Administration on the basis of false intelligence reports that Iraq had secret 'weapons of mass destruction' (Fawcett, 2013; Herring and Robinson, 2014–2015; Western, 2005). The 2003 invasion of Iraq is just one of the many conflicts affecting the lives of millions of people around the world as the United States and its allies attempt to control the development of global capitalism in a way that best suits their economic interests and the US desire to retain its status as the world's sole global hegemon. The geopolitical instability this engenders results in many needless deaths, much suffering and immense and long-lasting environmental damage.

Destructive material capabilities in an unstable world order: twenty-first-century military conflicts

The year 2003 heralded a significant development in global politics, with the United States bypassing the international legal requirement of obtaining UN Security Council authorisation for its proposed military action in Iraq (Franck, 2003, 2006; Kramer, Michalowski and Rothe, 2005).[14] The millions of people around the world protesting against military action were also ignored (Anderson, Bennis and Cavanagh, 2003; Anderson et al., 2003) and the invasion proceeded. The invasion of Iraq set off a chain of events that resulted in the deaths of 'tens if not hundreds of thousands' of Iraqi civilians (Fawcett, 2013) and the rapid rise of Islamic State (IS) (Arbatova and Dynkin, 2016), thereby triggering the massive (and ongoing) destabilisation of the Middle East (Barton, 2016; Ezrow, 2016; Fawcett, 2013; Newsinger, 2015). The long-term effects of the invasion of Iraq continue to unfold throughout the world in the form of terrorist attacks that are met by further repression by authorities, which leads to more terrorism and creates a vicious circle of escalating violence (Keane, 2015; Stern and McBride, 2013).

The armed conflicts that have subsequently erupted in the destabilised Middle East and North Africa (MENA) region have also contributed to the 'record-high' numbers of 'displaced' people worldwide (ICRC, 2016), many of them seeking refuge in Europe (UNHCR, 2016; WEF, 2016).[15] This unprecedented displacement of large numbers of people has prompted not only a series of refugee crises but also very dangerous nationalist, xenophobic and racist reactions from some European politicians and citizens (CIVICUS, 2016; Saull, 2015). Another effect of the spiral of violence is the loss of many fundamental civil rights that were previously taken for granted in Western liberal democratic societies (Cox cited in Martin, 2013; Jarvis and Lister, 2013), including the right to privacy; as revealed by Edward Snowden, ordinary people all over the world are now subjected to widespread and indiscriminate surveillance (Altschuler, 2015; Brevini 2015) publicly justified with reference to the 'War on Terror'.

The effects of the 2003 invasion of Iraq illustrate how the foreign policy of the United States and its allies, particularly in the MENA region, but also in their dealings with China and the Russian Federation, generates an increasingly unstable world order that further delegitimises global ruling elites in the eyes of many people – especially those living in the affected regions. The US-led attack on Iraq occurred in the context of the second Bush Administration's response to the September 2001 al-Qaeda terrorist attacks on US territory, which was to declare a 'War on Terror' (Krebs and Lobasz, 2007). This 'War on Terror' was to be conducted through what has come to be known as the 'Bush Doctrine', a policy characterised by unilateralism, 'pre-emptive war' and 'regime change' (Jervis, 2016) coupled with a violation of the Geneva Conventions.[16] Afghanistan was the first target of the 'War on Terror', and the US and its allies invaded the country in 2001, launching a long-term conflict that spilled over into Pakistan (Shaw, 2013; Roberts, 2009) and is still ongoing. This was followed by the illegal invasion of Iraq in 2003 (Kramer and Michalowski, 2005) and the arrest, show trial (Peterson, 2007) and execution of its president, Saddam Hussein. The next target of 'regime change' was Libyan president Muammar Gaddafi, who was killed by the US- and NATO-supported anti-government National Transitional Council in 2011 (Karniel, Lavie-Dinur and Azran, 2015).

US foreign policy is also involved in, or provides support to, a variety of other military attacks against IS, the successor to al-Qaeda, in countries such as Syria (Guerlain, 2014) and Yemen (Borger and Jacobs, 2017). These military actions have further destabilised the MENA region, killing and maiming an unknown number of innocent civilians and causing much ongoing suffering in the region (Crawford, 2013, 2016; Dehghan and Algohbary, 2017; Shaw, 2013). The instability is compounded by US support for Israeli enmity against Iran (Pillar, 2016), which is also a regional rival of another US ally, Saudi Arabia (Guerlain, 2014). Adding further complexities to the unstable MENA region is the US rivalry with China, which has become 'the second-largest trading partner in the Arab world, and the first trading partner of nine Arab states' as well as the Gulf's main oil client since 2014 (Kausch, 2015). US conflicts with Russia are also evident in the war in Syria, where Russia's only remaining external naval base is located in the Mediterranean port of Tartus, and where Russia has intervened against US efforts to achieve regime change (Kausch, 2015).

Critical GPE theorists argue that US confrontations with China and Russia are largely motivated by economic concerns, which Desai, Freeman and Kagarlitsky (2016, p. 498) describe as 'the real drivers of capitalist international relations'. Hudson (2016), for instance, argues that China's rising economic power is threatening to US interests in many ways.

> It provides other countries, rich as well as poor, with alternative trade and financial options. And it threatens US control over international economic governance, undermining the legitimacy of its power in existing institutions

via the deadlocked WTO... and the increasingly isolated IMF and World Bank. Building alternative economic institutions in the face of US and Western resistance reflects the emerging multipolar reality.

(Hudson, 2016, p. 557)

The abundant natural gas and fishery resources at stake in the maritime disputes in the South China Sea are additional economic factors informing the US challenge to China's dominance in this region (Morton, 2016). Along similar lines, the US is keen to ensure that US-based corporations acquire access to Russia's gas markets in Europe, where they intend to sell their shale gas (Chanis, 2012; Kausch, 2015; Ratner et al., 2015), and the provocation leading to Russia's annexation of the Crimea (Cypher, 2016) presented a perfect excuse for the US to press its European allies to stop importing natural gas from Russia (Davenport and Erlanger, 2014; Goldenberg, 2014).

In addition to these broader economic interests that contribute to US military interventions in various parts of the world, there are many other ways in which conflict benefits US corporations. The fact that US companies such as former Halliburton subsidiary KBR profit immensely from government contracts to build and maintain military bases (Vine, 2015) is also significant. These corporations are, moreover, not the only economic beneficiaries of conflict, with the sale of weaponry itself constituting a lucrative industry that generates billions of dollars in annual sales (Rufanges, 2016; Guay, 2015) and partly drives what Cypher (2016) refers to as the US governments' 'relentless pursuit of global militarism'.[17] War, and the tools of war that generate enormous profits, are not only an integral part of capitalist economies but also constitute yet another way in which wealth is transferred from taxpayers to private corporations (Vine, 2015) and a particularly perverse cause of unnecessary, severe and long-lasting ecological damage.

Material capabilities: the effects of capitalist militarism on the biosphere

In addition to the economic drain on scarce public resources that could be better spent on measures addressing climate change and on social programmes that advance the development of humanity, the many negative consequences of the current militarism, which is driven by capitalist rivalries, include the immense environmental damage that is caused both by the 'normal' operations of military bases (Vine, 2015) and by wars (Al-Azzawi, 2016; Kiernan, 2015; Lawrence et al., 2015; Mathieson, 2014), issues which security analysts generally ignore. Traditional IR theorists and security analysts limit their concerns regarding the environment to the way in which 'resource scarcity' and climate change can serve as 'threat multipliers' that can contribute to future conflicts, or to the way in which the effects of climate change might damage military infrastructure (Livingstone, 2015; Milman, 2016). In addition, while some traditional IR theorists and security analysts theorise about the possibility of nuclear war (Frühling and O'Neil, 2017), their concerns are generally confined to geostrategic issues and seldom extend to the environmental damage and human deaths and suffering that would result from such a conflict. Conversely, analysts of the social and environmental costs of wars and nuclear weapons fail to include discussions of the wider context within which these weapons are manufactured, tested, sold, and deployed; for example, while Kristensen and McKinzie (2015) discuss the environmental and social costs of detonating nuclear weapons, and Crowley and Ahearne (2002) discuss how the ensuing environmental damage can be 'managed', the authors of both papers neglect the role of the wider economic and geostrategic interests that constitute the underlying causes of these problems.

Problem-solving theory versus critical theory: different forms of analysis

It is not surprising that many (although not all) of the discipline experts whose work is cited in this chapter adopt a problem-solving approach, focusing narrowly on some issues while neglecting others. Ecosocialist John Bellamy Foster points out that:

> as a rule, the social sciences are compromised from the start. As shown in particular by the discipline of economics, they are ideologically compelled to answer all concrete issues in terms set by capitalism, excluding any perspective that seriously challenges that system or its boundaries. Social scientists are thus discouraged from questioning – or indeed even naming – the fundamental structures and workings of the historical system in which we live. It follows that the social-scientific contributions most relevant to our understanding of the causes and imperatives of climate change have originated outside the mainstream of academic social science, in critical analyses of capitalism.
>
> (Foster, 2017, p. 4)

In contrast to the problem-solving approaches dominating analyses of the geostrategic issues discussed above, ecosocialist Ian Angus (2016) situates his discussion of the causes and effects of global warming and climate change within the wider context of a critical analysis of capitalism. As part of his critique, Angus explains that the environmental impacts of detonating nuclear weapons, contrary to any notions that they can be 'managed', are so severe and so long-lasting that geologists working in the Anthropocene Working Group are considering 'Residues from hydrogen bomb explosions that began in 1952 [and] peaked in 1961–62, leaving a clear worldwide signature' (Angus, 2016, p. 57) as a potential 'stratigraphic signature' marking the end of the Holocene and the beginning of the Anthropocene. This reference to debates about the origins of the Anthropocene constitutes part of Angus' wider discussion that critically analyses capitalism as the underlying cause of the current and interrelated ecological, economic and socio-political crises – that is, in the context of a discussion that 'connects the dots' between these issues and thus better reflects the complexity of the real world than analyses that adopt single-issue problem-solving approaches. One of the greatest strengths of ecosocialist critical analyses of the causes of the current ecological, economic and socio-political crises is their ability to 'connect the dots' and illuminate the interrelationships between all the issues outlined in this chapter.

Notes

1 As noted in Chapter 2, however, it is important to keep in mind the critique that neo-Gramscians misinterpret what Gramsci meant when they overemphasise the consensual aspect of hegemony (for example, refer to Budd, 2013).

2 US political economist Frances Fukuyama wrote an influential article entitled 'The End of History?' which was published in a US foreign policy magazine, *The National Interest*, in 1989, and in which he notoriously declared that the collapse of the Soviet Union and the shift to capitalist economies and, more importantly, to Western consumer cultures in the former USSR and in China, signalled 'the unabashed victory of economic and political liberalism' (Fukuyama, 1989, p. 3).

3 In advanced capitalist economies, 'underdevelopment' is measured against 'development' using crude indicators such as gross domestic product (GDP) that fail to consider either the environmental costs or the distributional outcomes of increases in GDP.

4 The understanding that capitalism develops unevenly has been traced back to the writings of Marx and Engels, and this understanding also informed Lenin's analyses of imperialism (Löwy, 2010; Pröbsting, 2016). Desai, Freeman and Kagarlitsky (2016) provide a concise overview of the way in which uneven and combined development explains the expansion of global capitalism after imperialism's 'Thirty Years' Crisis' (1914–1945) that began with the outbreak of World War I.

5 Some scientists are, however, concerned that concentrations of methane may prove to be more significant than CO_2 concentrations. According to Canadell et al. (2016), methane emissions have been increasing in the past 20 years, with the concentration of this gas 'growing ten times faster' since 2007 than it did in the early 2000s, and increasing 'faster still in 2014 and 2015'. Anthropogenic sources of methane emissions include agricultural activities and the mining and use of fossil fuels such as coal, oil and natural gas (Canadell et al., 2016).

6 The energy imbalance is caused by more energy entering the Earth's atmosphere than is re-radiated into space as a result of the GHGs trapping the incoming heat.

7 'We Are Not All in This Together' is the title of Chapter 11 in Angus' (2016) book *Facing the Anthropocene: Fossil Capitalism and the Crisis of the Earth System*.

8 According to Rockström (2015, p. 5), four out of nine planetary boundaries have been transgressed: biosphere integrity, interference with the nitrogen and phosphorous cycles, climate change and land use change.

9 The reserve army of labour is a Marxist concept and refers to the ranks of currently unemployed workers whose mere existence gives employers greater bargaining power when determining the wages and working conditions for those fortunate enough to be employed.

10 While Piketty has been widely praised for his empirical research, the theoretical basis of his analysis (particularly the way in which he defines 'capital' and the way his analysis omits any considerations of class) and the solutions he proposes have

been critiqued by many analysts (for example, Peet, 2015; Reitz, 2016; Thompson, 2014). It is also important to note that Piketty (2014) largely ignores ecological economics in his brief discussion on climate change, which is confined to a few observations about how climate change will negatively affect GDP and a few statements on the importance of protecting 'natural capital'.

11 Widespread and persistent protests against the Icelandic government's acceptance of IMF conditions for a loan needed to bail out the banks that crashed as a result of the GFC resulted in the government's resignation and the election of a new government that temporarily nationalised the banks and implemented measures that reduced household and non-financial business debts (Hart-Landsberg, 2013).

12 As many political analysts note, however, it is important to distinguish between different forms of populism; while Donald Trump in the US, Marine le Pen in France and Pauline Hanson in Australia represent right-wing populist politicians who advocate socially damaging policies, some Left-leaning politicians such as Jeremy Corbyn in the UK, Bernie Sanders in the US and the late Venezuelan president Hugo Cháves are often similarly derided in the media as 'populist' while, in reality, their policies represent progressive ideals (albeit to varying degrees).

13 Estimates of how many people participated in these protests vary: Hil (2008) reports an estimated total of 10 million people in over 800 cities, while Carty (2009) reports an estimated total of 15 million people in 75 cities. As McPhail and McCarthy (2004) point out, attendance numbers at such mass-protest events are often disputed, with authorities tending to downplay the numbers and organisers and supporters tending to exaggerate them.

14 US policymakers decided to bypass the UN procedures because France, Russia and Germany opposed war with Iraq (Kramer and Michalowski, 2005), and a Security Council Resolution authorising it was likely to be vetoed (Anderson, Bennis and Cavanagh, 2003; Anderson et al., 2003).

15 According to a United Nations High Commissioner for Refugees report, by the end of 2015 '65.3 million individuals were forcibly displaced worldwide as a result of persecution, conflict, generalized violence, or human rights violations. This is 5.8 million more than the previous year (59.5 million)' (UNHCR, 2016, p. 2). The total number of refugees at the end of 2015 was estimated to be 16.1 million people, 'the highest level in the past two decades and approximately 1.7 million [people] more than the total reported 12 months earlier' (UNHCR, 2016, p. 13).

16 The Geneva Conventions are part of a body of international law that stipulates that 'parties to armed conflict protect civilians and non-combatants, limit the means or methods that are permissible during warfare and conform to rules governing the behaviour of occupying forces' (Kramer and Michalowski, 2005, p. 451) and some legal experts argue that the Bush Administration's failure to conform to these stipulations in the 2003 Iraq war constitutes a state crime (Kramer and Michalowski, 2005; Kramer, Michalowski and Rothe, 2005).

17 Six American corporations, which rank amongst the top eight firms with earnings based on 'defense-related revenues', dominate the global defense industry (Guay, 2015).

References

Al-Azzawi, S.N. (2016). The Deterioration of Environmental and Life Quality Parameters in Iraq Since the 2003 American Occupation of Iraq. *International Journal of Contemporary Iraqi Studies*, 10(1, 2), 53–72.

Alpher, D. (2017). Charlottesville and the Politics of Fear. The Conversation, 17 August. Available at https://theconversation.com/charlottesville-and-the-politics-of-fear-82443.

Altschuler, B.E. (2015). Is the Pentagon Papers Case Relevant in the Age of WikiLeaks? *Political Science Quarterly*, 130(3), 401–423.

Anderson, S., Bennis, P. and Cavanagh, J. (2003). *Coalition of the Willing or Coalition of the Coerced? How the Bush Administration Influences Allies in its War on Iraq*. Institute for Policy Studies. Available from www.ips-dc.org/wp-content/uploads/2003/02/COERCED.pdf.

Angus, I. (2016). *Facing the Anthropocene: Fossil Capitalism and the Crisis of the Earth System*. New York: Monthly Review Press.

Arbatova, N.K. and Dynkin, A.A. (2016). World Order after Ukraine. *Survival: Global Politics and Strategy*, 58(1), 71–90.

Arctic Council (2016). *Arctic Resilience Report*. Carson, M. and Peterson, G. (eds). Stockholm Environment Institute and Stockholm Resilience Centre. Available at www.sei-international.org/mediamanager/documents/Publications/ArcticResilience Report-2016.pdf.

Ashman, S. (2009). Capitalism, Uneven and Combined Development and the Transhistoric. *Cambridge Review of International Affairs*, 22(1), 29–46.

Aykut, S.C. (2016). Taking a Wider View on Climate Governance: Moving beyond the 'Iceberg,' the 'Elephant,' and the 'Forest'. *WIREs Climate Change*, 7, 318–328.

Barton, G. (2016). Out of the Ashes of Afghanistan and Iraq: The Rise and Rise of Islamic State. The Conversation, 2 March. Available at https://theconversation.com/out-of-the-ashes-of-afghanistan-and-iraq-the-rise-and-rise-of-islamic-state-55437.

Bieler, A. (2012). 'Workers of the World, Unite'? Globalisation and the Quest for Transnational Solidarity. *Globalizations*, 9(3), 365–378.

Borger, J. and Jacobs, B. (2017). Yemen Wants US to Reassess Counter-Terrorism Strategy after Botched Raid. *The Guardian*, 8 February. Available at www.theguardian.com/world/2017/feb/08/yemen-us-raid-al-qaida-counter-terrorism-strategy-trump.

Brands, H. and Feaver, P. (2016). Stress-Testing American Grand Strategy. *Survival*, 58(6), 93–120.

Brevini, B. (2015). Western Democracy's New Maxim: Surveillance and Soft Despotism. The Conversation, 18 December. Available at https://theconversation.com/western-democracys-new-maxim-surveillance-and-soft-despotism-48879.

Bruff, I. (2014). The Rise of Authoritarian Neoliberalism. *Rethinking Marxism*, 26(1), 113–129.

Budd, A. (2013). *Class, States and International Relations: A Critical Appraisal of Robert Cox and Neo-Gramscian Theory*. London: Routledge.

Campbell, E. (2013). Typhoon Haiyan Exposes Different Perspectives on a Warming Planet. Australian Broadcasting Corporation News, 25 November. Available at www.abc.net.au/news/2013-11-26/climate-change-science-debate-continues-amid-haiyan-recovery/5115456.

Canadell, P., Poulter, B., Saunois, M., Krummel, P., Bousquet, P. and Jackson, R. (2016). Methane from Food Production Might be the Next Wildcard in Climate Change. The Conversation, 12 December. Available at https://theconversation.com/methane-from-food-production-might-be-the-next-wildcard-in-climate-change-69894.

Caraus, T. and Parvu, C.A. (2017). Cosmopolitanism and Global Protests. *Globaliza-tions*, 14(5), 659–666.

Carroll, T. and Jarvis, D.S.L. (2015). The New Politics of Development: Citizens, Civil Society, and the Evolution of Neoliberal Development Policy. *Globalizations*, 12(3), 281–304.

Carty, V. (2009). The Anti-War Movement versus the War Against Iraq. *International Journal of Peace Studies*, 14(1), 17–38.

Chadwick, R., Good, P., Martin, G. and Rowell, D.P. (2016). Large Rainfall Changes Consistently Projected over Substantial Areas of Tropical Land. *Nature Climate Change*, 6, 177–181.

Chanis, J. (2012). U.S. Liquefied Natural Gas Exports and America's Foreign Policy Interests. *American Foreign Policy Interests*, 34(6), 329–334.

Chou, M. (2015). From Crisis to Crisis: Democracy, Crisis and the Occupy Move-ment. *Political Studies Review*, 13(1), 46–58.

CIVICUS (2016). *State of Civil Society Report 2016*. Available at http://civicus.org/index.php/socs2016.

Cox, R.W. (1981). Social Forces, States, and World Orders: Beyond Inter-national Relations Theory. *Millennium: Journal of International Studies*, 10(2), 126–155.

Cox, R.W. (1983). Gramsci, Hegemony and International Relations: An Essay in Method. *Millennium: Journal of International Studies*, 12(2), 162–175.

Cox, R.W. (1987). *Production, Power and World Order: Social Forces in the Making of History*. New York: Columbia University Press.

Cox, R.W. (1996b). Towards a Posthegemonic Conceptualization of World Order: Reflections on the Relevancy of Ibn Khaldun (1992). In Cox, R.W. with Sinclair, T.J., *Approaches to World Order*. Cambridge: Cambridge University Press, 144–173.

Cox R.W. with Schechter, M.G. (2002). *The Political Economy of a Plural World: Crit-ical Reflections on Power, Morals and Civilization*. New York: Routledge.

Crawford, N.C. (2013). *Civilian Death and Injury in the Iraq War, 2003–2013*. Watson Institute for International and Public Affairs, Brown University. Available at http://watson.brown.edu/costsofwar/files/cow/imce/papers/2013/Civilian%20Death%20and%20Injury%20in%20the%20Iraq%20War%2C%202003-2013.pdf.

Crawford, N.C. (2016). *Update on the Human Costs of War for Afghanistan and Pakistan, 2001 to mid-2016*. Watson Institute for International and Public Affairs, Brown University. Available at http://watson.brown.edu/costsofwar/files/cow/imce/papers/2016/War%20in%20Afghanistan%20and%20Pakistan%20UPDATE_FINAL_corrected%20date.pdf.

Cronin, B. (2013). Some 95% of 2009–2012 Income Gains Went to Wealthiest 1%. *The Wall Street Journal*, 10 September. Available at http://blogs.wsj.com/economics/2013/09/10/some-95-of-2009-2012-income-gains-went-to-wealthiest-1/.

Crowley, K.D. and Ahearne, J.F. (2002). Managing the Environmental Legacy of U.S. Nuclear-Weapons Production: Although the Waste from America's Arms Buildup Will Never be 'Cleaned Up,' Human and Environmental Risks Can be Reduced and Managed. *American Scientist*, 90(6), 514–523.

Cypher, J.M. (2016). Hegemony, Military Power Projection and US Structural Eco-nomic Interests in the Periphery. *Third World Quarterly*, 37(5), 800–817.

Davenport, C. and Erlanger, S. (2014). US Hopes Boom in Natural Gas Can Curb Putin. *The New York Times*, 5 March. Available at www.nytimes.com/2014/03/06/world/europe/us-seeks-to-reduce-ukraines-reliance-on-russia-for-natural-gas.html?_r=0.

Dehghan, S.K. and Algohbary, A. (2017). Yemen's Food Crisis: 'We Die Either from the Bombing or the Hunger'. *The Guardian*, 8 February. Available from www.theguardian.com/global-development/2017/feb/08/yemen-food-crisis-we-are-broken-bombing-hunger.

della Porta, D. (2012). Mobilizing Against the Crisis, Mobilizing for 'Another Democracy': Comparing two Global Waves of Protest. *Interface*, 4(1), 274–277.

Desai, R., Freeman, A. and Kagarlitsky, B. (2016). The Conflict in Ukraine and Contemporary Imperialism. *International Critical Thought*, 6(4), 489–512.

di Muzio, T. (2015). The Plutonomy of the 1%: Dominant Ownership and Conspicuous Consumption in the New Guilded Age. *Millennium: Journal of International Studies*, 43(2), 492–510.

Ezrow, N. (2016). The Catastrophic Legacy of 9/11 Will Define the US for Years to Come. The Conversation, 9 September. Available at https://theconversation.com/the-catastrophic-legacy-of-9-11-will-define-the-us-for-years-to-come-64067.

Fandos, N. (2017). Climate March Draws Thousands of Protesters Alarmed by Trump's Environmental Agenda. *The New York Times*, 29 April. Available at www.nytimes.com/2017/04/29/us/politics/peoples-climate-march-trump.html.

Fawcett, L. (2013). The Iraq War Ten Years On: Assessing the Fallout. *International Affairs*, 89(2), 325–343.

Foster, J.B. (2017). Trump and Climate Catastrophe. *Monthly Review*, 68(9), 1–17.

Foster, J.B., Clark, B. and York, R. (2010). *The Ecological Rift: Capitalism's War on the Earth*. New York: Monthly Review Press.

Fouskas, V.K. (2016). How Class and the Rise of China Won Trump the White House. The Conversation, 6 December. Available at https://theconversation.com/how-class-and-the-rise-of-china-won-trump-the-white-house-69515.

Franck, T.M. (2003). What Happens Now? The United Nations after Iraq. *The American Journal of International Law*, 97(3), 607–620.

Franck, T.M. (2006). The Power of Legitimacy and the Legitimacy of Power: International Law in an Age of Power Disequilibrium. *The American Journal of International Law*, 100(1), 88–106.

Frühling, S. and O'Neil, A. (2017). Nuclear Weapons and Alliance Institutions in the Era of President Trump. *Contemporary Security Policy*, 38(1), 47–53.

Fukuyama, F. (1989). The End of History? *The National Interest*, 16, Summer 1989, 3–18.

Gallagher, K.P. (2016). Trade, Investment, and Climate Policy: The Need for Coherence. In Gallagher, K.P. and Barakatt, C. (eds), *Trade in the Balance: Reconciling Trade and Climate Policy: Report of the Working Group on Trade, Investment, and Climate Policy*. The Frederick S. Pardee Center for the Study of the Longer-Range Future and the Global Economic Governance Initiative, Boston University. Available at www.bu.edu/pardee/2016/11/03/new-report-trade-in-the-balance-reconciling-trade-and-climate-policy/.

Gilens, M. and Page, B.I. (2014). Testing Theories of American Politics: Elites, Interest Groups, and Average Citizens. *American Political Science Association*, 12(3), 564–581.

Goldenberg, S. (2014). US Expands Gas Exports in Bid to Punish Putin for Crimea. *The Guardian*, 25 March. Available at www.theguardian.com/environment/2014/mar/25/us-expands-gas-exports-in-bid-to-punish-putin-for-crimea.

Guay, T. (2015). US Remains Top Arms Exporter, but Russia is Nipping at its Heels. The Conversation, 24 March. Available at https://theconversation.com/us-remains-top-arms-exporter-but-russia-is-nipping-at-its-heels-38639.

Guerlain, P. (2014). Obama's Foreign Policy: 'Smart Power,' Realism and Cynicism. *Society*, 51(5), 482–491.

Hansen, J., Sato, M., Hearty, P., Ruedy, R., Kelley, M., Masson-Delmotte, V., Russell, G., Tselioudis, G., Cao, J., Rignot, E., Velicogna, I., Kandiano, E., von Schuckmann, K., Kharecha, P., Legrande, A.N., Bauer, M. and Lo, K-W. (2016). Ice Melt, Sea Level Rise and Superstorms: Evidence from Paleoclimate Data, Climate Modelling, and Modern Observations that 2°C Global Warming Could be Dangerous. *Atmospheric Chemistry and Physics*, 16, 3761–3812.

Harrison, P.A., Dunford, R.W., Homan, I.P. and Rousevell, M.D.A. (2016). Climate Change Impact Modelling Needs to Include Cross-Sectoral Interactions. *Nature Climate* Change, 6, 885–892.

Hart-Landsberg, M. (2013). Lessons from Iceland: Capitalism, Crisis, and Resistance. *Monthly Review*, 65(5), 26–44.

Harvey, D. (2010). *The Enigma of Capital and the Crises of Capitalism*. New York: Oxford University Press.

Helleiner, E. (2013). Did the Financial Crisis Generate a Fourth Pillar of Global Economic Architecture? *Swiss Political Science Review*, 19(4), 558–563.

Herring, E. and Robinson, P. (2014–2015). Report X Marks the Spot: The British Government's Deceptive Dossier on Iraq and WMD. *Political Science Quarterly*, 129(4), 551–583.

Heyes, J., Lewis, P. and Clark, I. (2012). Varieties of Capitalism, Neoliberalism and the Economic Crisis of 2008–?. *Industrial Relations Journal*, 43(3), 222–241.

Hil, R. (2008). Civil Society, Public Protest and the Invasion of Iraq. *Social Alternatives*, 27(1), 29–33.

Hudson, M. (2016). Ukraine and the New Economic Cold War. *International Critical Thought*, 6(4), 556–569.

Intergovernmental Panel on Climate Change (IPCC) (2014). *Climate Change 2014: Synthesis Report*. Contribution of Working Groups I, II and III to the Fifth Assessment Report of the Intergovernmental Panel on Climate Change. Core Writing Team, Pachauri, R.K. and Meyer, L.A. (eds). IPCC, Geneva. Available at https://epic.awi.de/id/eprint/37530/.

Intergovernmental Panel on Climate Change (IPCC) (2019a). Summary for Policymakers. In Shukla, P.R., Skea, J., Calvo Buendia, E., Masson-Delmotte, V., Pörtner, H-O., Roberts, D.C., Zhai, P., Slade, R., Connors, S., van Diemen, R., Ferrat, M., Haughey, E., Luz, S., Neogi, S., Pathak, M., Petzold, J., Portugal Pereira, J., Vyas, P., Huntley, E., Kissick, K., Belkacemi, M., Malley, J. (eds), *Climate Change and Land: An IPCC Special Report on Climate Change, Desertification, Land Degradation, Sustainable Land Management, Food Security, and Greenhouse Gas Fluxes in Terrestrial Ecosystems*. Available at www.ipcc.ch/srccl/.

Intergovernmental Panel on Climate Change (IPCC) (2019b). Summary for Policymakers. In Pörtner, H-O., Roberts, D.C., Masson-Delmotte, V., Zhai, P., Tignor,

M., Poloczanska, E., Mintenbeck, K., Alegría, A., Nicolai, M., Okem, A., Petzold, J., Rama, B., Weyer, N.M. (eds), *IPCC Special Report on the Ocean and Cryosphere in a Changing Climate*. Available at www.ipcc.ch/srocc/.

Internal Displacement Monitoring Centre (IDMC) (2015). Global Estimates 2015: People Displaced by Disasters. Available at www.internal-displacement.org/publications/2015/global-estimates-2015-people-displaced-by-disasters.

International Committee of the Red Cross (ICRC) (2016). *Protracted Conflict and Humanitarian Action: Some Recent ICRC Experiences*. International Committee of the Red Cross, Geneva. Available at www.icrc.org/sites/default/files/document/file_list/protracted_conflict_and_humanitarian_action_icrc_report_lr_29.08.16.pdf.

Jamieson, A. (2017). After the Women's March: Six Mass US Demonstrations to Join this Spring. *The Guardian*, 5 February. Available at www.theguardian.com/world/2017/feb/05/womens-march-mass-protests-scientists-immigrants-climate-change.

Jarvis, L. and Lister, M. (2013). Disconnected Citizenship? The Impacts of Anti-Terrorism Policy on Citizenship in the UK. *Political Studies*, 61(3), 656–675.

Jefferies, W. (2015). On the Alleged Stagnation of Capitalism. *Review of Radical Political Economics*, 47(4), 588–607.

Jervis, R. (2016). Understanding the Bush Doctrine: Preventive Wars and Regime Change. *Political Science Quarterly*, 131(2), 285–311.

Karanikolos, M., Mladovsky, P., Cylus, J., Thomson, S., Basu, S., Stuckler, D., Mackenbach, J.P. and McKee, M. (2013). Financial Crisis, Austerity, and Health in Europe. *The Lancet*, 381(9874), 1323–1331.

Karniel, Y., Lavie-Dinur, A. and Azran, T. (2015). Broadcast Coverage of Gaddafi's Final Hours in Images and Headlines: A Brutal Lynch or the Desired Death of a Terrorist? *The International Communication Gazette*, 77(2), 171–188.

Kausch, K. (2015). Competitive Multipolarity in the Middle East. *The International Spectator*, 50(3), 1–15.

Keane, J. (2013). A Short History of Banks and Democracy. The Conversation, 22 April. Available at https://theconversation.com/a-short-history-of-banks-and-democracy-11991.

Keane, J. (2015). War Comes Home. The Conversation, 15 November. Available at https://theconversation.com/war-comes-home-50715.

Keaney, M. (2014). Financialization and Social Structures of Accumulation Theory. *World Review of Political Economy*, 5(1), 45–77.

Kiernan, K. (2015). Nature, Severity and Persistence of Geomorphological Damage Caused by Armed Conflict. *Land Degradation & Development*, 26, 380–396.

Klein, N. (2014). *This Changes Everything: Capitalism vs the Planet*. London: Allen Lane.

Kramer, R.C. and Michalowski, R.J. (2005). War, Aggression and State crime: A Criminological Analysis of the Invasion and Occupation of Iraq. *British Journal of Criminology*, 45(4), 446–469.

Kramer, R., Michalowski, R. and Rothe, D. (2005). 'The Supreme International Crime': How the U.S. War in Iraq Threatens the Rule of Law. *Social Justice*, 32(2), 52–81.

Krebs, R.R. and Lobasz, J.K. (2007). Fixing the Meaning of 9/11: Hegemony, Coercion, and the Road to War in Iraq. *Security Studies*, 16(3), 409–451.

Kristensen, H.M. and McKinzie, M.G. (2015). Nuclear Arsenals: Current Developments, Trends and Capabilities. *International Review of the Red Cross*, 97(899), 563–599.

Lapavitsas, C. (2013). The Financialisation of Capitalism: 'Profiting without Producing'. *City*, 17(6), 792–805.

Lapavitsas, C. (2014). Turn this Ship Around: Confronting Financialised Capitalism. *Juncture*, 21(1), 35–39.

Lapavitsas, C. and Mendieta-Muñoz, I. (2016). The Profits of Financialization. *Monthly Review*, 68(3), 49–62.

Lawrence, M.J., Stemberger, H.L.J., Zolderdo, A.J., Struthers, D.P. and Cooke, S.J. (2015). The Effects of Modern War and Military Activities on Biodiversity and the Environment. *Environmental Reviews*, 23(4), 443–460.

Leichenko, R. and Silva, J.A. (2014). Climate Change and Poverty: Vulnerability, Impacts, and Alleviation Strategies. *WIREs Climate Change*, 5, 539–556.

Liu, W., Xie, S., Liu, Z. and Zhu, J. (2017). Overlooked Possibility of a Collapsed Atlantic Meridional Overturning Circulation in Warming Climate. *Science Advances*, 3(1), e1601666.

Livingstone, D.N. (2015). The Climate of War: Violence, Warfare, and Climatic Reductionism. *WIREs Climate Change*, 6, 437–444.

Löwy, M. (2010). *The Politics of Combined and Uneven Development*. Chicago, IL: Haymarket Books.

Magdoff, F. (2015). A Rational Agriculture is Incompatible with Capitalism. *Monthly Review*, 66(10), 1–18.

Martin, S.J. (2013). In Conversation with Robert Cox: Historical Change, the Occupy Movement and Frozen Social Forces. *Global Social Policy*, 13(2), 216–225.

Mathiesen, K. (2014). What's the Environmental Impact of Modern War? *The Guardian*, 6 November. Available at www.theguardian.com/environment/2014/nov/06/whats-the-environmental-impact-of-modern-war.

Matthijs, M. (2014). Mediterranean Blues: The Crisis in Southern Europe. *Journal of Democracy*, 25(1), 101–115.

Mcphail, C. and McCarthy, J. (2004). Who Counts and How: Estimating the Size of Protests. *Contexts*, 3(3), 12–18.

Milman, O. (2015). James Hansen, Father of Climate Change Awareness, Calls Paris Talks 'A Fraud'. *The Guardian*, 12 December. Available at www.theguardian.com/environment/2015/dec/12/james-hansen-climate-change-paris-talks-fraud.

Milman, O. (2016). Military Experts Say Climate Change Poses 'Significant Risk' to Security. *The Guardian*, 14 September. Available at www.theguardian.com/environment/2016/sep/14/military-experts-climate-change-significant-security-risk.

Milman, O. (2017). March for Science Puts Earth Day Focus on Global Opposition to Trump. *The Guardian*, 22 April. Available at www.theguardian.com/environment/2017/apr/22/march-for-science-earth-day-climate-change-trump.

Morton, K. (2016). China's Ambition in the South China Sea: Is a Legitimate Maritime Order Possible? *International Affairs*, 92(4), 909–940.

Newsinger, J. (2015). Wars Past and Wars to Come. *Monthly Review*, 67(6), 34–40.

Norris, J.R., Allen, R.J., Evan, A.T., Zelinka, M.D., O'Dell, C.W. and Klein, S.A. (2016). Evidence for Climate Change in the Satellite Cloud Record. *Nature*, 536(7614), 72–75.

Nuccitelli, D. (2017). We're Now Breaking Temperature Records Once Every Three Years. *The Guardian*, 23 January. Available at www.theguardian.com/environment/climate-consensus-97-per-cent/2017/jan/23/were-now-breaking-global-temperature-records-once-every-three-years.

O'Connor, J. (1989). Uneven and Combined Development and Ecological Crisis: A Theoretical Introduction. *Race & Class*, 30(3), 1–11.

Oltermann, P. (2017). Protesters Plan to 'Kettle' Leaders at G20 Summit in Hamburg. *The Guardian*, 3 July. Available at www.theguardian.com/world/2017/jul/03/protesters-plan-to-kettle-leaders-at-g20-summit-in-hamburg.

Ortiz, I., Burke, S., Berrada, M. and Cortés, H. (2013). *World Protests 2006–2013*. Initiative for Policy Dialogue and Friedrich-Ebert-Stiftung New York Working Paper 274. Available at http://policydialogue.org/publications/working_papers/world_protests_2006-2013/.

Oxfam (2017). *An Economy for the 99%, Oxfam Briefing Paper*. Available at www.oxfam.org/sites/www.oxfam.org/files/file_attachments/bp-economy-for-99-percent-160117-en.pdf.

Peck, J. (2013). Explaining (with) Neoliberalism. *Territory, Politics, Governance*, 1(2), 123–157.

Peet, R. (2015). Capital in the 21st Century: Economics as Usual. *Geoforum*, 65, 301–303.

Peterson, J. (2007). Unpacking Show Trials: Situating the Trial of Saddam Hussein. *Harvard International Law Journal*, 48(1), 257–292.

Piketty, T. (2014). *Capital in the Twenty-First Century*. Cambridge, MA: Belknap Press.

Pillar, P.R. (2016). The Role of Villain: Iran and U.S. Foreign Policy. *Political Science Quarterly*, 131(2), 365–385.

Price, A. (2016). The 'Left-Behind' Once Had a Real Voice: The Globalisation Protesters of the 1990s. The Conversation, 24 November. Available at https://theconversation.com/the-left-behind-once-had-a-real-voice-the-globalisation-protesters-of-the-1990s-68940.

Pröbsting, M. (2016). Capitalism Today and the Law of Uneven Development: The Marxist Tradition and its Application in the Present Historic Period. *Critique: Journal of Socialist Theory*, 44(4), 381–418.

Ramsey, J.G. (2012). Revolution Underground? Critical Reflections on the Prospect of Renewing Occupation, Socialism and Democracy. *Socialism and Democracy*, 26(3), 93–116.

Ratner, M., Parfomak, P.W., Luther, L. and Fergusson, I.F. (2015). *U.S. Natural Gas Exports: New Opportunities, Uncertain Outcomes*. Congressional Research Service. Available at www.ourenergypolicy.org/u-s-natural-gas-exports-new-opportunities-uncertain-outcomes-3/.

Reitz, C. (2016). Accounting for Inequality: Questioning Piketty on National Income Accounts and the Capital-Labor Split. *Review of Radical Political Economists*, 48(2), 310–321.

Roberts, A. (2009). Doctrine and Reality in Afghanistan. *Survival*, 51(1), 29–60.

Rockström, J. (2015). *Bounding the Planetary Future: Why We Need a Great Transition*. Great Transition Initiative: Towards a Transformative Vision and Praxis. Available at www.greattransition.org/images/GTI_publications/Rockstrom-Bounding_the_Planetary_Future.pdf.

Rockström, J., Schellnhuber, H.J., Hoskins, B., Ramanathan, V., Schlosser, P., Brasseur, G.P., Gaffney, O., Nobre, C., Meinshausen, M., Rogelj, J. and Lucht, W. (2016). The World's Biggest Gamble. *Earth's Future*, 4, 465–470.

Rufanges, J.C. (2016). The Arms Industry Lobby in Europe. *American Behavioral Scientist*, 60(3), 305–320.

Saull R. (2015). Capitalism, Crisis and the Far-Right in the Neoliberal Era. *Journal of International Relations and Development*, 18(1), 25–51.

Schlosberg, D. and Collins, L.B. (2014). From Environmental to Climate Justice: Climate Change and the Discourse of Environmental Justice. *WIREs Climate Change*, 5, 359–374.

Schouten, P. (2009). Theory Talk #37: Robert Cox on World Orders, Historical Change, and the Purpose of Theory in International Relations. Theory Talks. Available at www.theory-talks.org/2010/03/theory-talks-37.html.

Sharples, J. (2016). Firestorms: The Bushfire/Thunderstorm Hybrids We Urgently Need to Understand. The Conversation, 10 November. Available at https://theconversation.com/firestorms-the-bushfire-thunderstorm-hybrids-we-urgently-need-to-understand-68426.

Shaw, I.G.R. (2013). Predator Empire: The Geopolitics of US Drone Warfare. *Geopolitics*, 18(3), 536–559.

Shearer, C., Chio, N., Myllyvirta, L., Ye, A. and Nace, T. (2016). *Boom and Bust 2016: Tracking the Global Coal Plant Pipeline*. Coalswarm, Sierra Club and Greenpeace. Available at http://sierraclub.org/sites/www.sierraclub.org/files/uploads-wysiwig/final%20boom%20and%20bust%202017%20(3-27-16).pdf.

Steffen, W., Grineveld, J., Crutzen, P. and McNeill, J. (2011). The Anthropocene: Conceptual and Historical Perspectives. *Philosophical Transactions of the Royal Society*, 369, 842–867.

Steffen, W., Rockström, J., Richardson, K., Lenton, T.M., Folke, C., Liverman, D., Summerhayes, P., Barnosky, A., Cornell, S.E., Crucifix, M., Donges., J.F., Fetzer, I., Lade, S.J., Scheffer, M., Winkelmann, R., Schellnhuber, H.J. (eds) (2018). Trajectories of the Earth System in the Anthropocene. *Proceedings of the National Academy of Sciences*, 115(33), 8252–8259.

Stern, J. and McBride, M.K. (2013). *Terrorism after the 2003 Invasion of Iraq*. Watson Institute for International and Public Affairs, Brown University. Available at http://watson.brown.edu/costsofwar/files/cow/imce/papers/2013/Terrorism%20after%20the%202003%20Invasion%20of%20Iraq.pdf.

Stevens, M. (2017). The Tax March Explained: Protesters Hope to Pressure Trump into Releasing Returns. *The New York Times*, 15 April. Available at www.nytimes.com/2017/04/15/us/politics/tax-day-march.html.

Tanuro, D. (2013). *Green Capitalism: Why it Can't Work*. London: The Merlin Press Ltd.

Thompson, M.J. (2014). Mapping the New Oligarchy. *New Politics*, 15(1), 125–130.

United Nations Conference on Trade and Development (UNCTAD) (2016). *Trade and Development Report, 2016*. UNCTAD, Geneva. Available at http://unctad.org/en/pages/PublicationWebflyer.aspx?publicationid=1610.

United Nations Environment Programme (UNEP) (2016). *Resource Efficiency: Potential and Economic Implications*. UNEP International Resource Panel Report. Available at www.unep.org/resourcepanel/.

United Nations Environment Programme (UNEP) (2017). *The Status of Climate Change Litigation: A Global Review*. UNEP and Columbia Law School. Available at http://wedocs.unep.org/bitstream/handle/20.500.11822/20767/climate-change-litigation.pdf?sequence=1&isAllowed=y.

United Nations High Commissioner for Refugees (UNHCR) (2016). *Global Trends: Forced Displacement in 2015*. Available at https://s3.amazonaws.com/unhcrsharedmedia/ 2016/2016-06-20-global-trends/2016-06-14-Global-Trends-2015.pdf.

Vine, D. (2015). The United States Probably Has More Foreign Military Bases than Any Other People, Nation, or Empire in History. *The Nation*, 14 September. Available at www.thenation.com/article/the-united-states-probably-has-more-foreign-military-bases-than-any-other-people-nation-or-empire-in-history.

Waddock, S. (2016). Neoliberalism's Failure Means We Need a New Narrative to Guide Global Economy. The Conversation, 6 December. Available at https://the conversation.com/neoliberalisms-failure-means-we-need-a-new-narrative-to-guide-global-economy-69096.

Western, J. (2005). The War over Iraq: Selling War to the American Public. *Security Studies*, 14(1), 106–139.

Weston, D. (2014). *The Political Economy of Global Warming: The Terminal Crisis*. New York: Routledge.

Whiteford, P. (2017). Do 8 Men Really Control the Same Wealth as the Poorest Half of the Global Population? The Conversation, 18 January. Available at https:// theconversation.com/do-8-men-really-control-the-same-wealth-as-the-poorest-half-of-the-global-population-71406.

Widmaier, W. (2016). Trump's Carrier Coup Reveals Credibility Gap between Twitter Rhetoric and Economic Reality. The Conversation, 1 December. Available at https://theconversation.com/trumps-carrier-coup-reveals-credibility-gap-between-twitter-rhetoric-and-economic-reality-69573.

Williams, J. and Crutzen, P.J. (2013). Perspectives On Our Planet in the Anthropocene. *Environmental Chemistry*, 10(4), 269–280.

Witte, G. (2015). Greeks Emphatically Reject Austerity, Elect Syriza in Historic Vote. *The Washington Post*, 25 January. Available at www.washingtonpost.com/world/ greeks-expected-to-elect-leftist-syriza-party-in-sunday-vote/2015/01/25/6e273 dea-a246-11e4-91fc-7dff95a14458_story.html.

World Bank (2014). *Turn Down the Heat: Confronting the New Climate Normal*. Available at www-wds.worldbank.org/external/default/WDSContentServer/WDSP/ IB/2014/11/20/000406484_20141120090713/Rendered/PDF/927040v20WP00 O0ull0Report000English.pdf.

World Economic Forum (WEF) (2016). *The Global Risks Report 2016* (11th edn). Available at http://reports.weforum.org/global-risks-2016/.

World Economic Forum (WEF) (2020). *The Global Risks Report 2020* (15th edn). Available at www.weforum.org/reports/the-global-risks-report-2020.

World Health Organisation (WHO) (2016). *Preventing Disease through Healthy Environments: A Global Assessment of the Burden of Disease from Environmental Risks*. Available at http://un.org.au/2016/03/16/preventing-disease-through-healthy-environments-a-global-assessment-of-the-burden-of-disease-from-environmental-risks/.

World Meteorological Organization (WMO) (2019). *WMO Statement on the State of the Global Climate in 2018*. WMO-No. 1233. Available at https://public.wmo.int/ en/resources/library/wmo-statement-state-of-global-climate-2018.

Youngs, R. (2017). We Live in a Global 'Age of Rage' – and it's Entering a New Phase. The Conversation, 19 July. Available at https://theconversation.com/we-live-in-a-global-age-of-rage-and-its-entering-a-new-phase-80173.

8 Prospects for climate justice

A research agenda

Introduction

Challenging dominant IR perspectives whose purpose it is to preserve the status quo, the philosophy of 'responsible scholarship' underlying Cox's MHS promotes a critical research agenda that can be placed 'at the service of the weak' by envisioning alternative futures and supporting transformative, emancipatory change (Falk, 2016). As Sinclair (1996, p. 15) puts it, Cox's academic work 'contains an emancipatory project which means that it should be of the highest priority to anyone wishing to understand the world in order to change it'. In addition to these acknowledgements of how Robert Cox reminded GPE theorists that '[t]heory is always *for* someone and *for* some purpose' (Cox, 1996, p. 87; emphases in original), the importance of his specific conceptual contributions to GPE scholarship are also widely acknowledged; for example, he is credited with introducing 'the application of Gramscian concepts at the international level' (Gill, 1993, p. 4) and initiating a 'crucial break … from mainstream International Relations (IR) approaches to hegemony' (Bieler and Morton, 2004, p. 85; see also Morton, 2006), as well as with developing a 'deeply historical and relational approach' that challenges 'the very core of the discipline of International Relations' (Brincat, 2016, p. 506).

The utility of an ecological neo-Gramscian perspective in the Anthropocene

In this book the modified, ecological version of the critical GPE neo-Gramscian perspective encloses human reproductive and productive capabilities within the confines of the Earth's biosphere and is used to analyse the origins and ongoing causes of global capital's early-twenty-first-century organic crisis (refer to Figure 8.1). One aim of the extended case study presented in this book was to demonstrate the power of Robert Cox's Method of Historical Structures (MHS) as an heuristic device which, in modified form, provides an analytical tool that can accommodate the complex analytical challenges of the Anthropocene. This is particularly the case if the ecological neo-Gramscian perspective is used in conjunction with the most powerful tool

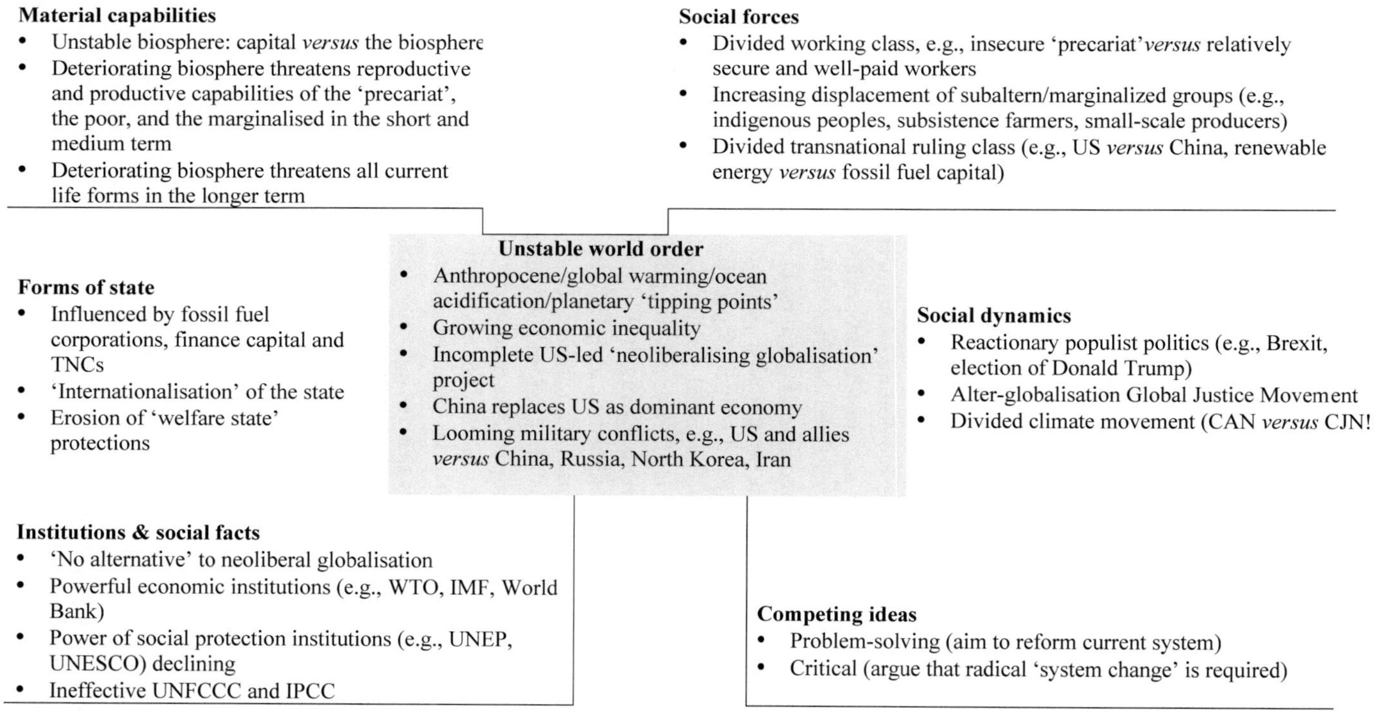

Material capabilities
- Unstable biosphere: capital *versus* the biosphere
- Deteriorating biosphere threatens reproductive and productive capabilities of the 'precariat', the poor, and the marginalised in the short and medium term
- Deteriorating biosphere threatens all current life forms in the longer term

Forms of state
- Influenced by fossil fuel corporations, finance capital and TNCs
- 'Internationalisation' of the state
- Erosion of 'welfare state' protections

Institutions & social facts
- 'No alternative' to neoliberal globalisation
- Powerful economic institutions (e.g., WTO, IMF, World Bank)
- Power of social protection institutions (e.g., UNEP, UNESCO) declining
- Ineffective UNFCCC and IPCC

Unstable world order
- Anthropocene/global warming/ocean acidification/planetary 'tipping points'
- Growing economic inequality
- Incomplete US-led 'neoliberalising globalisation' project
- China replaces US as dominant economy
- Looming military conflicts, e.g., US and allies *versus* China, Russia, North Korea, Iran

Social forces
- Divided working class, e.g., insecure 'precariat' *versus* relatively secure and well-paid workers
- Increasing displacement of subaltern/marginalized groups (e.g., indigenous peoples, subsistence farmers, small-scale producers)
- Divided transnational ruling class (e.g., US *versus* China, renewable energy *versus* fossil fuel capital)

Social dynamics
- Reactionary populist politics (e.g., Brexit, election of Donald Trump)
- Alter-globalisation Global Justice Movement
- Divided climate movement (CAN *versus* CJN!

Competing ideas
- Problem-solving (aim to reform current system)
- Critical (argue that radical 'system change' is required)

Figure 8.1 Overview of extended case study: an ecological neo-Gramscian concept map of the global political economy of climate change.

available for analysing the inner workings of global capital: the Marxist critical perspective.

Like many other studies of the current climate change crisis, the evidence presented in this book demonstrates that because of capital's normative assumptions that the effects of human activities on the biosphere can be discounted or completely ignored, the human species is at a very dangerous point in its history. It is imperative that we change the way we understand our place within the Earth System, and our dependence on its continued functioning in a way that supports current life-forms that inhabit our planet. Understanding is not, however, enough; it must also be translated into appropriate action. Since theory informs action, universities and other institutions of learning have a central role to play in this respect. As Clive Hamilton argues, we need to be teaching theory that is appropriate for the Anthropocene.

A GPE research agenda for climate justice in the Anthropocene

Concrete examples of important climate justice issues that could be the subject of further research using the ecological neo-Gramscian perspective (or any other theoretical lens appropriate for the Anthropocene) include identifying and finding solutions to the threats faced by those least responsible for, and most threatened by, the unfolding ecological disasters. For example, it is important to understand the way in which the global expansion and deepening of capitalist relations of production disrupts the lives of people with relatively little formal power, such as indigenous peoples and small-scale farmers (Dell'Angelo et al., 2017), and how the resulting challenges are compounded not only by the effects of anthropogenic global warming (Connolly-Boutin and Smit, 2016) but also by the way in which policies designed to mitigate climate change further threaten their reproductive and productive capabilities (Dehm, 2016). Critical research projects could investigate the strengths and weaknesses of counter-hegemonic movements resisting these threats. Also important is research investigating how workers' calls for a 'just transition' to a decarbonised economy (Hampton, 2018) could be supported in ways that lead towards the eventual institutionalisation of socio-economic arrangements that are not only ecologically sustainable but also fairer and more genuinely democratic.

It is imperative that researchers who support the aims of climate justice engage in an emerging issue of potentially catastrophic consequences: plans to 'geo-engineer' the planet in order to 'buy time' for reducing greenhouse gas emissions at some future date. The possible economic, political, cultural, ethical and even existential implications for different socio-economic classes, indigenous peoples, subsistence farmers and future generations if plans to 'manage' and delay the effects of continuing GHG emissions by using large-scale 'geo-engineering' technologies are implemented are very grave. Such plans are currently gaining momentum (Corry, 2017; Huttunen, Skytén and Hilden, 2015) despite the risks that they may fail, or could themselves generate new ecological catastrophes either now or for future generations (for discussions of such concerns, refer to Keller, Feng and Oschlies, 2014 and McCormack et al., 2016). A growing community of scholars aims to develop governance mechanisms for the implementation of 'solar radiation management' (for example, refer to Larson, 2016 and Nicholson, Jinnah and Gillespie, 2018) – a form of problem-solving theory that could ultimately be used to legitimise such large-scale, inherently undemocratic technological 'solutions' (Harnish, Uther and Boettcher, 2015). Critical theorists wanting to support those who lack the power to participate in making decisions about possible future trajectories of our shared planet could engage in research that seeks ways of avoiding such large-scale experimentation with the Earth System, which is a chaotic, non-linear dynamic system that is not subject to human control, by working collaboratively with communities and groups to change the socio-economic systems that people *do* have the power to shape.

In conclusion, taking the implications of the Anthropocene seriously entails a recognition that traditional IR and IPE concerns such as 'hegemony' (in the orthodox Realist meaning of powerful states), the state of the global economy and 'security' need to be folded into, and subsumed by, focused explorations of how we can create socially just human systems that protect the biosphere and ensure a habitable planet for both present and future generations of people and for all other life-forms. It is in this way that the responsible scholarship that Robert Cox advocates could be conducted with the purpose of serving the least powerful groups, and the other life-forms that we share the planet with, rather than the interests of the most powerful, who have proved themselves incapable of finding effective solutions to the crises they have caused and continue to exacerbate. It is this inability of those who have power to resist the independent dynamics of capital as self-expanding value that leads ecosocialists to argue that real solutions to the ecological breakdown we face will require system change. The subtitle of the Climate & Capitalism website clearly and concisely states the ecosocialist case: 'Ecosocialism or barbarism: There is no third way'.

References

Bieler, A. and Morton, A.D. (2004). A Critical Theory Route to Hegemony, World Order and Historical Change: Neo-Gramscian Perspectives in International Relations. *Capital & Class*, 28(1), 85–113.

Brincat, S.K. (2016). From International Relations to World Civilizations: The Contributions of Robert W. Cox. *Globalizations*, 13(5), 506–509.

Connolly-Boutin, L. and Smit, B. (2016). Climate Change, Food Security, and Livelihoods in Sub-Saharan Africa. *Regional Environmental Change*, 16, 385–399.

Corry, O. (2017). The International Politics of Geoengineering: The Feasibility of Plan B for Tackling Climate Change. *Security Dialogue*, 48(4), 297–315.

Cox, R.W. (1996). Social Forces, States, and World Orders: Beyond International Relations Theory (1981). In Cox, R.W. with Sinclair, T.J., *Approaches to World Order*. Cambridge: Cambridge University Press, 85–123.

Dehm, J. (2016). Indigenous Peoples and REDD+ Safeguards: Rights as Resistance or as Disciplinary Inclusion in the Green Economy? *Journal of Human Rights and the Environment*, 7(2), 170–217.

Dell'Angelo, J., D'Odorico, P., Rulli, M.C. and Marchand, P. (2017). The Tragedy of the Grabbed Commons: Coercion and Dispossession in the Global Land Rush. *World Development*, 92, 1–12.

Falk, R. (2016). On the Legacy of Robert W. Cox. *Globalizations*, 13(5), 501–505.

Gill, S. (ed.) (1993). *Gramsci, Historical Materialism and International Relations*. Cambridge: Cambridge University Press.

Hampton, P. (2018). Trade Unions and Climate Politics: Prisoners of Neoliberalism or Swords of Climate Justice? *Globalizations*, 15(4), 470–486.

Harnish, S., Uther, S. and Boettcher, M. (2015). From 'Go Slow' to 'Gung Ho'?: Climate Engineering Discourses in the UK, the US, and Germany. *Global Environmental Politics*, 15(2), 57–78.

Huttunen, S., Skytén, S. and Hilden, M. (2015). Emerging Policy Perspectives on Geoengineering: An International Comparison. *The Anthropocene Review*, 2(1), 14–32.

Keller, D.P., Feng, E.Y. and Oschlies, A. (2014). Potential Climate Engineering Effectiveness and Side Effects During a High Carbon Dioxide-Emission Scenario. *Nature Communications*, 5, 3304.

Larson, E.J. (2016). The Red Dawn of Geoengineering: First Step toward an Effective Governance of Stratospheric Injections. *Duke Law & Technology Review*, 14(1), 157–191.

McCormack, C.G., Born, W., Irvine, P.J., Achterberg, E.P., Amano, T., Ardron, J., Foster, P.N., Gattuso, J-P., Hawkins, S.J., Hendy, E., Kissling, W.D., Lluch-Cota, S.E., Murphy, E.J., Ostle, N., Owens, N.J.P., Perry, R.I., Pörtner, H.O., Scholes, R.J., Schurr, F.M., Schweiger, O., Settele, J., Smith, R.K., Smith, S., Thompson, J., Tittensor, D.P., van Kleunen, M., Vivian, C., Vohland, K., Warren, R., Watkinson, A.R., Widdicombe, S., Williamson, P., Woods, E., Blackstock, J.J. and Sutherland, W.J. (2016). Key Impacts of Climate Engineering on Biodiversity and Ecosystems, with Priorities for Future Research. *Journal of Integrative Environmental Sciences*, 13(2–4), 103–128.

Morton, A.D. (2006). The Grimly Comic Riddle of Hegemony in IPE: Where is Class Struggle? *Politics*, 21(6), 62–72.

Nicholson, S., Jinnah, S. and Gillespie, A. (2018). Solar Radiation Management: A Proposal for Immediate Polycentric Governance. *Climate Policy*, 18(3), 322–334.

Sinclair, T.J. (1996). Beyond International Relations Theory: Robert W. Cox and Approaches to World Order. In Cox, R.W. with Sinclair, T.J., *Approaches to World Order*. Cambridge: Cambridge University Press, 3–18.

Index

Page numbers in *italics* denote figures.

Advisory Group on Greenhouse Gases (AGGG): establishment of 87–89; US government's role in disbanding 88–91, 103n6; *see also* United States (US)

alter-globalisation movement *see* Global Justice Movement (GJM)

Alternatiba 150, 153, 155–156; *see also* Coalition Climat 21; Conference of the Parties (COP), COP-21 (Paris, 2015); Paris2015; prefiguration

alternative climate change summits 121–122, 150, 153; *see also* climate justice movement

alternative futures 33, 37, 39–40, 47, 61–65, 75–76, 125n4, 132, 137, 138, 151–156, 192; *see also* counter-hegemonic bloc

Angus, I. 3–4, 54, 73, 76–77, 78n13, 81, 98, 131, 136, 151, 168, 180–181

Anthropocene 4–7, 26, 76–77, 165–167, 169, 180: complexity and 14–16, 43, 48, 165, 192–195; controversies about 4–7, 18n3, 71–73, 78n8; geological marker of 6–7; radical implications of for social sciences 2–3, 17–18, 31, 40–43, 192–195; as rupture 17, 151; *see also* Earth System; Great Acceleration

anthropogenic global warming (AGW) 2, 7, 12, 14; and capital accumulation 67–68, 147–148; ecological effects of 15, 43, 48, 67–68, 75–76, 167–170; evidence of 40, 82–84, 87–88, 92–95, 102, 103n7, 103n10, 110, 167–170; mitigation of 111; official responses to 81–91; uncertainties regarding 82; *see also* Intergovernmental Panel on Climate Change (IPCC); United Nations Framework Convention on Climate Change (UNFCCC); United States (US)

anti-critiques 58, 77n2; *see also* ecosocialism; second-stage ecosocialist theory

Arrhenius, S. 82

Australia: climate change denialism 112; extreme weather 18n2; greenhouse gas emissions 103n7; megafires 1, 14–16; official responses to AGW 84

Benton, T. 55, 60

biosphere: as ecological MHS category 2–4, 12, 14, 30–31, 38, 40–43, 48, 86, 118, 163–168, 192–194

Blockadia 138, 153; *see also* counter-hegemonic bloc; protests

blue-green alliance 165; *see also* green capitalism

Bond, P. 140, 148, 157n6

Breakthrough Institute 54, 70–71, 78n7; *see also* ecological modernisation

Brecher, J. 145, 156n3

Bretton Woods system 8–10, 12, 13, 31

Brundtland, G. 86

Budd, A. 29, 44, 49n6, 181n6

Burkett, P. 24, 53, 55, 57, 58, 59–66, 77n4, 77n5

Canada: greenhouse gas emissions of 103n7; official responses to AGW 83–84

capital: accumulation 10, 28–31, 54, 61, 65, 71, 74–75, 100, 147–148, 157n8, 167; definition of 26–27; green faction

capital: *continued*
    of 102, 120, 147–148, 157n8;
    primitive accumulation 28–29, 49n4,
    75, 119; and nature 65–69, 71–75,
    147, 166–167; as self-expanding value
    28–31; US capital 8–12, 31, 67, 76,
    83, 86–93, 101–102, 115, 157n8,
    164–165, 177–180, 195; *see also*
    capitalism; fossil capital; labour theory
    of value; imperialism; Marxism; value
capitalism: and accumulation crises 12,
    65–69, 115, 121; alternatives to 47,
    61–64, 76, 132–133, 137, 138, 151,
    153–156; and commodity production
    9, 27–28, 32, 49n3, 49n4, 59, 67,
    73–74, 76, 78n6, 100, 120–121, 148,
    171; and competition 8–10, 12, 29, 44,
    165, 177–180; definitions of 26–27;
    ecological damage of 24, 41, 65–69,
    73–76, 86, 98, 132, 151, 164–168,
    194; and economic growth 5–8, 12,
    14, 29, 31, 70, 71, 86, 102, 112,
    117–118, 121, 124n1, 147, 157n8,
    167, 173; and environmental relations
    14, 24, 30–31, 48, 54–55, 62, 66–72,
    75–76, 83–87, 90–92, 98–99, 101,
    111–112, 114–119, 123, 131–132,
    144–148, 163–170, 176–177, 179–180,
    181n3; and exploitation 27, 41–42,
    65–66, 72, 132, 144–147, 164–166;
    and greenhouse gas emissions 14,
    29–31, 61, 69, 83–85, 87, 89, 97–102,
    103n7, 104n15, 111, 120, 125n9,
    139–140, 142, 166, 181n5, 194–195;
    internal contradictions of 30–31, 41,
    45, 65–71, 138; and liberal democracy
    10–11, 44, 83–85, 172–176, 177; and
    military conflict 164–165, 177–180,
    181n4; monopoly capitalism 29, 76;
    organic crisis of 48, 53, 54, 123–124,
    133, 144, 147, 150, 151, 164, 174,
    192; potential to reform 77n3, 84–85,
    117–118, 139–143, 147–148; social
    relations of production of 24, 26–28,
    33, 98, 116, 143–144, 154, 168–173;
    see also class and class struggle; disaster
    capitalism; finance capital; Global
    Financial Crisis (GFC); green
    capitalism; Marxism; means of
    production; mode of production;
    uneven and combined development;
    value
carbon dioxide (CO_2) emissions *see*
    greenhouse gas (GHG) emissions

carbon trading: ecosocialist debates about
    139–143; ineffectiveness of 67, 69,
    100, 140–141, 143; *see also* false
    solutions
Carson, R. 85, 113
Cartesian dualism 70–73
Castree, N. 70
catastrophe bonds *see* weather derivatives
centrally planned economies 10; *see also*
    Cold War; ideology
China 11, 83, 99, 103n7, 164, 171,
    178–179
civil disobedience *see* protests
Clark, B. 57–61, 64–65, 73–75, 76,
    137–138, 157n5
class and class struggle 25–29, 39, 44–48,
    49n5, 49n7, 55, 56, 59–60, 62, 68–69,
    77n3, 118, 119, 121–124, 130, 133,
    137, 138, 140–148, 157n7, 164,
    168–176, 181–2n10, 194–195; *see also*
    class consciousness; trade unions;
    working class
class consciousness 157n7; Gramsci's
    levels of 47; *see also* trade unions;
    working class
Clean Development Mechanism (CDM)
    67, 102; *see also* false solutions
climate activism 117, 120, 132, 138, 139,
    153: Flood Wall Street action 125n3;
    Paris2015 148–150; People's Climate
    March 117
Climate Action Network International
    (CAN-I): achievements of 123; history
    of 120–121; radical critiques of 121;
    reformist aims of 120–121; *see also* false
    solutions
climate change: denialism 93, 112;
    negotiations 2, 81–82, 99–102, 103n5,
    113, 115, 123, 124n1, 125n6, 125n9,
    148–149; *see also* anthropogenic global
    warming (AGW); Conference of the
    Parties (COP); Intergovernmental
    Panel on Climate Change (IPCC)
climate crisis 1–2, 14, 16, 24, 27, 31, 33,
    55, 69, 75–76, 84–87, 111, 116, 118,
    120, 124, 132, 141, 142, 192–195;
    need for cooperation 164–165
climate justice: geo-engineering threats
    to 147–148, 157n9, 194–195; liberal
    interpretations of 99; US
    administration responses to 99–101,
    104n15; *see also* climate justice
    movement; Climate Justice Now!
    (CJN!); ecosocialism; false solutions;

just transition; real solutions; system change

climate justice movement 26, 54, 75, 76, 77, 81, 116–119; *see also* Climate Justice Now! (CJN!); ecosocialism; false solutions; system change

Climate Justice Now! (CJN!): ethical focus of 123; ideas defining 119, 122–124; origins of 121–122, 131; *see also* ecosocialism; false solutions; Global Justice Movement (GJM); system change

climate movement: climate action wing 2, 120–121; composition of 2, 116–117, 124n1, 134; ecosocialism and 130–131, 134, 136–143; ideological divisions within 102, 117–119, 122–124, 125n2, 125n3, 134, 135; origins of 110, 113, 116, 121–151; radical climate justice wing 2, 43, 54, 75, 77, 116–119, 121–123; war of position within 117–119, 134; *see also* Climate Action Network International (CAN-I); Climate Justice Now! (CJN!); ecosocialism

climate scientists: political naivety of 98; psychological stress of 98, 104n13

'climategate' *see* Intergovernmental Panel on Climate Change (IPCC)

Club of Rome 76, 85

Coalition Climat 21, 148–150, 153–156; *see also* Alternatiba; Conference of the Parties (COP), COP-21 (Paris, 2015); Paris2015; red lines protest

Cochabamba People's Agreement 122

Cold War 8, 10–11, 19n10; *see also* global political economy; ideology

commodity *see* capitalism

commodification: of carbon 120–121, 148; of nature 67, 120–121, 148; neoliberalism and 32, 67; *see also* green capitalism; neoliberalism

common but differentiated responsibilities (CBDR) 99, 101, 104n15; *see also* United Nations Framework Convention on Climate Change (UNFCCC); United States (US)

common sense 47–48, 118–119, 156; *see also* class consciousness; Gramsci, A.

communism 19n6; *see also* Cold War; ideology

comparative advantage 9–10

complex systems 16; obstacles to analyses of 16–18

competing ideas 37–38, 53–57, 87–88, 93–98, 130–139, 142, 151, 152; *see also* anti-critiques; ecosocialism; first-stage ecosocialist theory; Intergovernmental Panel on Climate Change (IPCC); Method of Historical Structures (MHS); second-stage ecosocialist theory; war of position

Conference of Polluters 149; *see also* Coalition Climat 21; Conference of the Parties (COP), COP-21 (Paris, 2015)

Conference of the Parties (COP): COP-1 (Berlin, 1995) 101; COP-6 (The Hague, 2000) 122; COP-13 (Bali, 2007) 121, 131; COP-15 (Copenhagen, 2009) 81, 95, 102, 121, 122, 123, 125n6; COP-21 (Paris, 2015) 102, 130, 148–155; and green capitalism 148; *see also* alternative climate change summits; climate movement; United Nations Framework Convention on Climate Change (UNFCCC)

conjunctural crisis 45, 174; *see also* organic crisis

Copenhagen Accord 123, 125n6; *see also* Conference of the Parties (COP); United Nations Framework Convention on Climate Change (UNFCCC); United States (US)

consumption 29, 54, 63–64, 103n7; *see also* capital; capitalism; value

co-optation *see* trasformismo

counter-hegemonic bloc 45–47, 119, 133, 138, 144–147, 151, 194–195; *see also* class consciousness; ecosocialism; Gramsci, A.; trasformismo; war of position

Cox, R.W.: distinction between critical theory and problem-solving theory 33–34, 39, 125n2; and emancipatory projects 33, 39–40, 192, 195; neo-Gramscian theoretical perspective 2–3, 33, 34–37, 39–40, 43, 114; responsible scholarship 192, 195; role of theory 33–34, 39–40, 192, 195; war of position 46–47; *see also* critical research agenda; hegemony; Method of Historical Structures (MHS)

critical research agenda 33, 192–195

critical theory 33–34, 39–40, 43–44, 53, 118, 119, 133, 134, 136, 178, 180–181, 192–195; *see also* Cox, R.W.; ecosocialism; Marxism

Daly, H. 61, 77n3
Davenport, N. 139–143, 142, 146; *see also* ecosocialism; false solutions; real solutions
de Lucia, V. 26, 119, 139, 147–148
democracy: as essentially contested concept 25; liberal 10–11, 172–174; and neoliberalism 174–176; as radical egalitarianism 54; *see also* Cold War; ideology
deregulation *see* neoliberalism
developing economies 9–10, 19n9, 87, 96, 99–102, 103n7, 115, 124n1, 137, 171, 181n3; *see also* Global South; uneven and combined development
dialectical analysis 58, 60, 70, 72–73, 77n2; *see also* ecosocialism; Marxism
Dickens, P. 57–58
disaster capitalism 66–68; *see also* capitalism; commodification; green capitalism
drought 14–15, 18n2, 48, 67, 83, 110, 116, 168; *see also* extreme weather

Earth System: science 3–8, 6, 17–18, 40–42, 48, 71–72, 76–77, 78n13, 98, 163, 165, 167–169, 192–195; tipping elements in 40, 145, 168, 169; *see also* Anthropocene; metabolic rift analysis; Method of Historical Structures (MHS)
Eastern Bloc 8, 10, 19n6, 171; *see also* Cold War; ideology
Eckersley, R. 61
ecological crisis *see* anthropogenic global warming (AGW); climate crisis; organic crisis
ecological modernisation 54, 69–70, 78n7; critiques of 18n3, 70–73, 78n8, 117–118; *see also* Breakthrough Institute
ecological neo-Gramscian Method of Historical Structures 2–3, 40–43, 163, 192–195, 193; *see also* Cox, R.W.; Gramsci, A.; Marxism; Method of Historical Structures (MHS)
ecomodernism *see* ecological modernisation
economic crises 1, 12–13, 45, 65–69, 111–112, 174–176, 180–181; *see also* capitalism; disaster capitalism; Global Financial Crisis (GFC); green capitalism; organic crisis
economic growth *see* capitalism
economic institutions post-WWII *see* Bretton Woods system

ecosocialism: and anti-capitalism 24–25, 33, 41, 53–54, 56, 76, 83, 84, 98, 131–132, 140–141, 147–148, 170, 180–181, 195; in climate justice movement 2, 116–117, 124, 130–139, 169–170; and climate negotiations 148–150; critiques of ecomodernism 69–71, 73; definition of 54; and ethics 118, 119, 123; influence of 130–131, 150; and market mechanisms 139–143; and metabolic rift theory 73–76; and prefiguration 151–156; strategy and tactics (overview) 130–139, 152; strengths of 134, 150–151; theoretical debates (varieties of ecosocialism) 54–69; and *trasformismo* 134, 139, 140; and war of position in climate movement 117–120; and working class struggle 143–148; *see also* counter-hegemonic bloc; first-stage ecosocialist theory; second-stage ecosocialist theory; third-stage ecosocialist theory
ecosocialist coalitions *see* Ecosocialist International Network (EIN); System Change Not Climate Change (SCNCC)
elites 48, 118, 121, 123, 124n1, 164, 168, 174–176, 178; *see also* class and class struggle
emancipatory projects *see* Cox, R.W
embedded liberalism: definition of 10; *see also* Cold War; ideology
emissions *see* greenhouse gas emissions
emissions trading 67, 69, 100, 102, 103, 139–143; *see also* false solutions
Engels, F. 24, 26, 54–58, 61–64, 73, 78n9, 144, 181n4; *see also* ecosocialism; Marx, K.; Marxism
Enlightenment project 75, 79
environmental crises *see* anthropogenic global warming (AGW); climate crisis; oceans
environmental justice movement 116, 122–123, 132; *see also* climate justice movement
environmental movement: and American Empire 114–115; and climate movement 113–117; and 'green imperialism' 115; and liberal ideology 77n3; origins of 85, 103n3; *see also* Climate Action Network International (CAN-I); climate justice movement; ecosocialism; false solutions

essentially contested concepts: definition of 25; practical implications of contested meanings 25–26; values and 25–26; *see also* ideology

extreme weather: definition of 18n2; ecological effects of 1, 43; occurrence of 14, 48, 83, 110, 168; socio-economic effects of 1, 43, 48, 67, 156, 170; terminology associated with 15

false solutions: and dangers of *trasformismo* 121; ecosocialist opposition to 139–143, 148; definitions of 33, 119; justifications for 139–140; and power structures 122, 148; radical climate justice movement opposition to 123–124, 134, 148; *see also* climate justice; System Change Not Climate Change (SCNCC)

finance capital 13, 29, 32, 67, 78n6, 171, 173–175

first-stage ecosocialist theory 54–56: second-stage ecosocialist critiques of first-stage ecosocialist theory 56–69; *see also* ecosocialism

Foran, J. 138, 149, 157n11

Forgacs, D. 45, 48

forms of state 36, 40, 44, 81, 174–176; *see also* Method of Historical Structures (MHS)

fossil capital 29–30

fossil fuels: and anthropogenic climate change 82–84, 87, 99, 181n5; and capitalism 29, 84, 87, 99, 103n7, 141, 142, 144, 146–147, 166–167, 168; and climate change denial 93; climate negotiations and the fossil fuel lobby 84, 89, 91–93, 96–98, 101, 112; climate movement opposition to 2, 117–118, 120, 122–123, 135–136; and greenhouse gas emissions 82–84, 87, 99, 181n5; *see also* fossil capital; just transition

Foster, J.B. 24, 53, 55–61, 64–77, 77n2, 78n9, 137–138, 157n5, 180

Fourier, J. 82

framework for action 34–35, 40; *see also* Method of Historical Structures (MHS)

General Agreement on Tariffs and Trade (GATT) 9; *see also* Bretton Woods system

Geneva Conventions 178, 182n16; *see also* United States (US)

geo-engineering 18n3, 71, 147, 157n9, 194; *see also* false solutions

Global Financial Crisis (GFC) 121, 164, 171–172, 175; *see also* protests

Global Justice Movement (GJM) 39, 123, 131, 176; *see also* climate justice movement; protests

Global North 14, 85, 96, 97, 101, 104n14, 115, 123, 166, 170, 175

global political economy 1945–1970s features of 8–12, 1980s–present features of 12–14, 31–32, 66–68, *100*, 120–121, *193*; and Anthropocene 3–8, 14–15, 31, 43, 68, 75–77, 84, 87, 99, 163–172, 179–180, 194; and Cold War 8, 10–11, 7n19; and conflict 29, 164–165, 177–179; critiques of possibilities of reforming 77n3, 121, 123–124, 141, 167, 170; dynamic nature of 27, 31–43, 73, 141, *142*; as field of study 32–34; and inequality 155, 170–176, 194–195; and political legitimacy 175–177; *see also* capital; capitalism; finance capital; fossil capital; fossil fuels; green capitalism; neoliberalism; uneven and combined development; United States (US)

Global South 14, 97, 101, 104n14, 123, 131, 166, 170–171; *see also* developing economies

global warming *see* anthropogenic global warming (AGW)

globalisation *see* neoliberalism

good sense *see* class consciousness

Gorz, A. 55

Gramsci, A. 43–48, 49n5, 49n7, 117, 118, 121, 133, 144, 172, 174, 181n1, 192

Great Acceleration 5–7; *see also* Anthropocene

green capitalism: and economic growth 117, 120, 147, 148, 157n8; ecosocialist critiques of 61, 76, 120, 125n5, 147, 157n8; *see also* false solutions; green imperialism; *trasformismo*; war of position

green imperialism 115; *see also* environmental movement

Greenhouse Gangsters vs Climate Justice 122; *see also* false solutions; green capitalism; *trasformismo*; war of position

greenhouse gas (GHG) emissions *see* anthropogenic global warming (AGW) habitability of Earth: risks to 14, 40, 41, 134, 146, 147, 153, 167–168, 195; *see also* Anthropocene; anthropogenic global warming (AGW); climate change; climate crisis; Earth System

Hahnel, R. 139–143, 157n6

Hamilton, C. 2–3, 17, 40, 43, 71–72, 76–77, 78n13, 151, 194

Hansen, J. 40, 82–83, 110, 137

Heatwaves 15; *see also* extreme weather; megafires

hegemony: and current world order 39–40, 114–115, 147, 164–165; Gramsci's approach to 44, 45, 49n5, 181; neo-Gramscian approach to 43, 44–45, 164, 171, 181; and organic crises 45–46; traditional international relations approach to 192, 195; *see also* counter-hegemonic bloc

heuristic device 39–40, 192; *see also* Method of Historical Structures (MHS)

historical bloc 44–47, 147; *see also* counter-hegemonic bloc; hegemony

historical materialism 42, 58, 73, 75, 78n9

historical structures: balance of forces within 35, 193; contradictions within 35, 39, 40; dynamic features of 32–33, 34–35, 193; *see also* Cox, R.W.; Method of Historical Structures (MHS)

'hockey stick' graph controversy 93, 94–95; *see also* climate change; Intergovernmental Panel on Climate Change (IPCC)

Holleman, H. 55, 56, 57

Holocene 3, 4, 17, 31, 43, 75, 167, 180; *see also* Anthropocene; Earth System; Great Acceleration

Hoodwinked in the Hothouse: False Solutions to Climate Change 119; *see also* commodification; green capitalism; false solutions

Horkheimer, M. 57

Hornick, B. 138–139

Hudis, P. 62

Hyman, R. 143–144, 157n7; *see also* trade unions

ideas: as category of Method of Historical Structures (MHS Forces) 35; competing ideas as element of MHS Forces Redux 37–38; *see also* Method of Historical Structures (MHS)

ideology: and class consciousness 47–48, 144, 146; and climate change denialism 95, 112, 169; and climate movement 116–117, 119, 142; and Cold War 8, 10–11; and embedded liberalism 114; and essentially contested concepts 25–26, 56–57, 59; and individualism 47, 49n8, 111; neoliberal 102, 167, 173, 180; and praxis 125n3; *see also* Cold War; essentially contested concepts; war of position

IMF *see* International Monetary Fund

imperialism 29, 164, 181n4; ecological imperialism 58, 74–75; green imperialism 115

Indigenous Peoples' Guide: False Solutions to Climate Change 119; *see also* false solutions

individualism: as explanation for barriers to climate action 111–112; *see also* ideology; liberalism; neoliberalism

Intended Nationally Determined Contributions (INDCs): 102, 123, 125n9

inter-generational equity 86; *see also* climate justice

Intergovernmental Panel on Climate Change (IPCC): as alternative to Advisory Group on Greenhouse Gases (AGGG) 89–91, 103n6; assessment reports 92, 103–4n10; 'climategate' controversy 95; discrediting of 93–96; ecosocialist assessment of 98; government control mechanisms over 91–96; 'hockey stick' graph controversy 94–95; ideologies informing critiques of 93–98; liberal critiques of 96–98; manufactured controversies about 93–96; member governments' review of IPCC assessment reports 92–93, 95–96, 104n11; restricted mandate of 92–93; structure of 92; Summary for Policymakers 92; US influence within 91–92, 97, 104n14; Working Groups 92, 96–97; *see also* Advisory Group on Greenhouse Gases (AGGG); United Nations Framework Convention on Climate Change (UNFCCC)

international institutions 2, 9, 11–12, 13, 84–87, 99, 178–179; and US influence 19n4; *see also* Bretton Woods system

International Monetary Fund (IMF) 9, 11, 12, 19n9, 120, 178–179, 182n11; *see also* Bretton Woods system
International Monetary System (IMS) 9; *see also* Bretton Woods system
IPCC *see* Intergovernmental Panel on Climate Change

just transition 123, 144, 146, 147, 194

Kahle, T. 145–146
Keeling Curve 82
Keeling, D. 82
Keynesianism *see* Cold War; embedded liberalism; ideology
Klein, N. 66–68, 153, 157n13, 168–169; liberal critiques of 134–135, 137–138; ecosocialist responses to critiques 137–139; socialist critiques of 125n4, 138
Kovel, J. 54, 55, 58, 76, 82, 131, 154, 156n3, 157n15–158n15
Kuhn, T. 3

La Via Campesina 132
labour movement: and Marxism 55; weakening of 12–13, 47, 170–171; *see also* class and class struggle; trade unions
Labor Network for Sustainability 146
labour power: as commodity 26, 28, 49n3; *see also* labour theory of value
labour theory of value 27–28, 59–61
Landless Workers' Movement 132
Latour, B. 70–71
laws of physics 42, 163, 168, 169; *see also* Earth System
Lenin, V.I. 29, 144, 181n4
liberal democracy *see* democracy
Lipow, G. 157n6
Longo, S. 57, 76
Löwy, M. 76, 131, 156n1
Lukács, G. 76
Luxemburg, R. 55

Mandel, E. 64
market mechanisms: as solutions to climate change 67–69, 100, 101–102, 119, 122, 130, 139–143, 148; *see also* commodification; false solutions
Marx, K. 24–29, 30, 42, 49n2, 49n5, 54–66, 70, 72–75, 78n9, 138, 144, 181n9; *see also* Marxism
Marxism: and 'ecological deficiencies' of Marx's labour theory of value 59–61;

environmental critiques of 54–59; and Frankfurt School critical theorists 57; and green theory 54–59; ideological motivations of critiques of 32, 56; as inherently ecological 53, 56, 58; on Marx's failure to identify the 'second contradiction of capitalism' 65–69; neglect of ecological content in 55; and postmodernism 56; 'productivism' and views on the human conquest of nature 61–65; *see also* alternative futures; anti-critiques; capital; capitalism; ecosocialism; first-stage ecosocialist theory; labour theory of value; Marx, K.; metabolic rift analysis; second-stage ecosocialist theory; value
material capabilities: as conceptual category of the Method of Historical Structures (MHS) 35, 40; as conceptual category of the Method of Historical Structures (MHS) Redux II 42; *see also* ecological neo-Gramscian Method of Historical Structures
means of production 27; *see also* capital; capitalism; class and class struggle; Marx, K.; Marxism
megafires: Australia (2019/2020) 1, 14–16; definition of 15; *see also* extreme weather
Mészáros, I. 76, 78n12
metabolic rift analysis 58, 72–76; *see also* ecosocialism; second-stage ecosocialist theory; Marxism
methane emissions *see* anthropogenic global warming (AGW)
Method of Historical Structures (MHS); and the Anthropocene 3, 41–42, 163–180, 192–195; critiques of MHS and MHS Redux 37, 40–41; Cox's MHS 34–37; Cox's MHS Forces 35; Cox's MHS Spheres 36; ecological neo-Gramscian MHS 3, 41–42; MHS Forces Redux II 3, 41–42, 42; Sinclair's MHS Redux 37–39; Sinclair's MHS Forces Redux 37; Sinclair's MHS Spheres Redux 38; strengths of MHS and MHS Redux 39–40
Middle East and North Africa (MENA) region 177; destabilisation of 177–178; refugees and displaced peoples 176, 177
military bases: environmental damage of 180

military conflicts: and capitalist rivalries 178–179; environmental damage of 179–180; socio-economic damage of 177–178

mode of production: capitalist 27–29; definition of 27; *see also* capital; capitalism; Marx, K.; Marxism

moderate climate action movement *see* climate action movement; Climate Action Network International (CAN-I)

Montreal Protocol on Substances that Deplete the Ozone Layer 90

Mooney, P. 148

Moore, J. 18n3, 72–73

Morton, A. 43

NATO *see* North Atlantic Treaty Organisation

nature: intrinsic value of 59–61; *see also* value

neo-Gramscian (ism) *see* Method of Historical Structures (MHS)

neoliberalism: advance wave of 13; climate crisis and 88, 111, 120–121; contested definitions of 26, 31–32; and environmental destruction 67, 113; and green capitalism 120–121; key features of 10, 67, 112, 171, 173–177; project vs trend 31–32; as social relation 39

Nordhaus, T. 70, 71, 78n7

North Atlantic Treaty Organisation (NATO) 8, 19n7, 178; *see also* Cold War

nuclear weapons; capitalist militarism and 179–180; environmental damage of 3, 85; social effects of 85, 180

oceans: effects of anthropogenic global warming on 6, 7, 14, 41, 43, 76, 93, 103–4n10, 166, 167, 168

O'Connor, J. 55, 58, 65–69, 165–166

oil workers' strike (US 2015) 143–148; *see also* class and class struggle; class consciousness; labour movement; trade unions

organic crisis 39, 53, 54, 116, 118, 123–124, 133, 164–165, 174, 192–194; definition of 45; and potential for counter-hegemonic bloc 46, 48, 124, 144, 147, 150–151; and potential for passive revolution 147–148, 150

organic intellectuals 44, 46, 47, 48, 138, 172, 173; *see also* war of position

paradigm shift 17: in natural sciences 17–18; in social sciences 17–18, 32; need for in the Anthropocene 3, 14, 17; *see also* Anthropocene; Earth System, science; global political economy as field of study; Method of Historical Structures (MSH)

Paris2015 149–150, 154–156; *see also* Alternatiba; Coalition Climat 21; Conference of the Parties (COP), COP-21 (Paris, 2015); red lines protest

Paris Agreement (PA) 1, 101–102, 167

passive revolution 46–47, 121, 123

Pax Americana see United States (US)

People's Climate March *see* climate activism

Piketty, T. 171–174, 181–2n10

planetary boundaries *see* tipping points

populism 175, 182n12

post-capitalism: Marx's vision 62–65

postmodernism 72–73

prefiguration 151–156, 157–8n15; *see also* Alternatiba; Transition Town movement

problem-solving theory 33–34, 44, 45, 53, 117–119, 172, 180–181; *see also* Cox, R.W.; critical theory; geo-engineering

proletariat *see* working class

protests 49n8, 117, 120, 122, 138, 149–150, 164, 176–177, 182n11

radical climate justice movement *see* climate justice movement; Climate Justice Now! (CJN!)

Ravindranath, N.H. 93

real solutions 118, 121, 124, 130, 134, 142, 147, 195; *see also* system change

red-green alliance *see* ecosocialism

red lines protest 149–150; *see also* Coalition Climat 21; Conference of the Parties (COP), COP-21 (Paris, 2015); Paris2015; protests

Reducing Emissions from Deforestation and Forest Degradation (REDD+) *see* commodification; false solutions

reductionist analyses *see* complex systems

research agenda *see* critical research agenda

reserve army of labour 171, 181n9

responsible scholarship *see* Cox, R.W.

Russian Federation 11, 178

Saito, K. 57
Salon, P. 149
Schmidt, A. 57, 70, 78n7
second contradiction of capitalism 65:
    critique of 65–69
second-stage ecosocialist theory 53–55,
    57–75; *see also* ecosocialism
Security Council *see* United Nations
    Security Council (UNSC)
Shellenberger, M. 70, 71
Sinclair, T. 3, 34–42, 175, 192
sixth great extinction event 4
Smith, N. 70
social dynamics 38; *see also* Method of
    Historical Structures (MHS); social
    movements
social facts 37–38; *see also* Method of
    Historical Structures (MHS)
social forces 38; *see also* Method of
    Historical Structures (MHS)
Solidarity 133, 136–137, 153
Special Report on Global Warming of
    1.5C (SR15) 1–2, 103n10
Stockholm Conference 85; *see also*
    United Nations Conference on the
    Human Environment (UNCHE)
Stockholm Declaration 86
subaltern classes and groups 39, 44–47,
    49n7, 119, 124, 142
system change 2, 111–112, 116, 118,
    119, 131, 134, 136–144, 147, 155,
    153, 154, 194–195; *see also* climate
    justice; ecosocialism; System Change
    Not Climate Change (SCNCC)
System Change Not Climate Change
    (SCNCC) 131–137, 143–144, 146,
    149–150

Tanuro, D. 55, 58, 76
third-stage ecosocialist theory 75–77; *see
    also* ecosocialism
Thomas, P. 43, 49n5
tipping points 40, 168, 169; *see also* Earth
    System
Tolba, M. 88–90
Tokar, B. 76, 156n3
Toronto Declaration 189; *see also* United
    States (US)
trade unions 12, 46, 137, 143–148, 157n7,
    157n8; *see also* labour movement
Trade Unions for Energy Democracy
    (TUED) 146
Transition Town movement 153–154;
    see also prefiguration

transitional demands 137, 139–143; *see
    also* false solutions; system change
trasformismo 46–47, 121, 134, 139–140, 148
Trotsky, L. 165
Tyndall, J. 82

UNEP *see* United Nations Environment
    Program (UNEP)
uneven and combined development
    165–166, 170–171, 181n4; *see also*
    capital; capitalism
UNFCCC *see* United Nations
    Framework Convention on Climate
    Change (UNFCCC)
United Nations (UN): power
    differentials within 19n9; specialised
    agencies 87; *see also* international
    institutions
United Nations Environment Program
    (UNEP) 85, 88–91
United Nations Framework Convention
    on Climate Change (UNFCCC), US
    influence in shaping 99–102
United States (US): and *Pax Americana*
    40, 114, 164; conflicts and
    environmental damage 164–165,
    177–180; and green imperialism 115;
    influence in climate change
    negotiations 99–102; influence in the
    Intergovernmental Panel on Climate
    Change (IPCC) 91–93, 97; and
    Montreal Protocol on Substances that
    Deplete the Ozone Layer 90–91; and
    neoliberalism 12–14, 31, 66–69, 88,
    101–102, 164; official responses to
    anthropogenic global warming 83,
    86–87, 104n14, 104n15; power within
    UN institutions 9, 11, 19n9, 90; role
    in disbanding the Advisory Group on
    Greenhouse Gases (AGGG) 88–91,
    103n6; role in shaping post-WWII
    global political economy 8–14, 164;
    and world order 9, 12, 39–40, 88,
    164, 177, 178

value 27–29, 49n2, 49n3, 54, 59–61,
    64–65; *see also* capital; capitalism;
    ecosocialism; fossil capital; Marx, K.;
    Marxism
Vienna Convention for the Protection of
    the Ozone Layer 88, 91, 103n5
Villach Conference 88–90; *see also*
    Advisory Group on Greenhouse Gases
    (AGGG)

war of manoeuvre 46
war of position 46–48, 117–119, 133, 147
Watt, J. 30
wealth 60–61
weather derivatives 67
Weston, D. 75–76, 154, 157n14
wildfires *see* megafires
Williams, C. 76, 81, 134–136, 139, 140, 156n3

working class 26–27, 29, 32, 39, 47, 119, 130, 138, 141; *see also* class and class struggle; class consciousness; trade unions
world order 8–13, 33, 36–40, 45, 81, 147, 153, 163–165, 177–179; *see also* United States (US)

York, R. 57